WHY WAR?
An Inquiry into the Genetic and Social Sources of Human Warfare

George R. Pitman

dog ear
PUBLISHING

First published by Dog Ear Publishing
4011 Vincennes Road
Indianapolis, IN 46268
www.dogearpublishing.net

dog ear
PUBLISHING

ISBN: 978-1-4575-3548-2
Library of Congress Control Number: 2015915251

This book is printed on acid free paper.
Printed in the United States of America

To my wife, Patricia,
and to Karen and David

Table of Contents

List of Illustrations

Preface

This book is the result of almost a lifetime of a concern about why humans are so warlike. I remember the radio announcements of Mussolini's invasion of Ethiopia in 1935 when I was only eight years old, of the German Anschluss of Austria, the invasion of Czechoslovakia in 1938 and of the outbreak of World War II in 1939. In 1940, the son of a close friend of my family who had joined the Royal Canadian Air Force was shot down and killed over Germany, and I remember the traumatic effect it had on his father. I also remember marching around a church auditorium in 1942 after the Japanese attack on Pearl Harbor, singing "Onward Christian Soldiers." At the time, it seemed right to me that my religious faith should support my country in its war against Germany and Japan after Japan's perfidious attack on Pearl Harbor. I also wondered whether German children were doing the same thing.

While I was in high school in 1944 and 1945, I worked summers and after school in a laboratory that was developing a substitute for the rubber latex that was no longer available because the Japanese had cut off our supply of rubber latex from Indo-China and our armed forces needed a substitute to keep their vehicles moving. It was a wonderful experience for a sixteen year old who aspired to become a scientist. I bought war bonds with the money I earned which later helped pay some of my college expenses, but I had no moral compunctions about my contributions to the war effort.

After I completed my graduate work in physics, I went to work in the aerospace industry, developing guidance systems for my country's intercontinental missiles. I remember a picture in the office of one of my colleagues of missile warheads raining down on an enemy city, enveloping it in smoke and flame. It may have inspired him in his work, but it bothered me that I was helping to develop such weapons. Later I went to work for a firm that had a contract with the US Arms Control and Disarmament Agency and because I had an interest in international relations I became the manager of the company's arms control project, which led me to study international relations at UCLA. These experiences allowed me to assuage my feelings of guilt, and in 1968 I was able to join the Arms Control Agency, where I became the science advisor to the bureau that led the strategic arms control negotiations with the Soviet Union to reduce and eliminate the strategic nuclear weapons of both countries, as well as becoming the science advisor to the US delegations that negotiated the arms-control treaties with the Soviet Union.

During most of my career, however, I became deeply puzzled about why humans are so warlike so after I retired I decided to devote myself to answering that question. At first I approached the problem by studying the history of warfare, especially wars between primitive societies, and I read such works as Konrad Lorentz's *On Aggression*, as well as much of the anthropological literature on war such as Arthur Ferrill's *On the Origin of War* and Lawrence Keeley's *War before Civilization*, but while this literature described warfare, none of it seemed to address the fundamental reasons of why we humans fight wars.

I decided that I might find the answers to my questions in evolutionary biology and went back to the university to study biology, human evolution, and evolutionary anthropology. The answers to my question are contained in the pages that follow. Because war is ubiquitous among humans, its sources must be embedded in the human genome. It is the result of a complex of genes, no one of which is a sufficient cause of human warfare. But human behavior is more than the dictates of our genes. Sometimes, we are able to suppress our baser instincts and behave in ways that express our nobler instincts that promote the wellbeing of all mankind, but nevertheless we continue to fight wars between nations, ethnic groups and religions.

If everything in this book were correct and was the final answer to the question of why humans are so warlike, I could not have written it. I hope, however, that it is a contribution to the answers to this vexing question and that it can shed light on how the human propensity for war can be reduced. As in all science, there is seldom a final answer. Each new theory and discovery is only an incremental step and also raises new questions which need to be investigated. My only hope is that this is a contribution to improving our understanding of why we are so warlike.

George R. Pitman
September 2015

Acknowledgments

This book would not have been possible without the help of many colleagues, friends, and teachers I have known over the years, but I am especially indebted to Victor Alessi, Mary Gallagher, and Stanly Riveles, who have read my manuscript and have offered many helpful comments and suggestions. I am also indebted to Colonel Glen Trimmer (US Air Force, retired), from whom I learned much about what motivates soldiers to fight and die in battles and to Rob Jolles, who offered many helpful suggestions on the process of getting this book published.

Chapter 1

Introduction

Only the dead have seen the end of war.

George Santayana[1]

War is the third largest cause of human mortality, following only disease and old age. It is estimated that more than one hundred million people, or 2.6 percent of the world's population, were killed in wars during the twentieth century, and that 62 percent of those deaths were non-combatants.[2] Between 60 and 80 million people were killed during World War II, including between 38 and 56 million noncombatants. More than 2.5 million lives were lost during the Korean War. During the war in Iraq, between 112,000 and 123,000 civilians lost their lives, 4,491 US service personnel were killed, and more than 32,000 US military personnel were wounded, many of whom will spend the rest of their lives without their limbs or sight, or will suffer from posttraumatic stress disorder and other mental disorders as a result of their war experiences.

Nearly all contemporary wars are either insurgencies to overthrow an existing government or wars between ethnic or religious groups. Between 1945 and mid-2014, there were approximately 51 interstate conflicts each of which had more than 25 battle-related deaths per year, compared with more than 240 civil, religious, and ethnic conflicts, each of which had more than 25 fatalities each year, and in more than half of these conflicts there were over a thousand fatalities each year.[3] Only three of the 43 ongoing wars in September 2014 were interstate wars; the rest were insurgencies, civil or religious wars.

Over four million pepole lost their lives to genocide in Rwanda, Burundi, the Democratic Republic of the Congo, Chechnya and Bosnia-Herzegovina during the 1990s. The most egregious genocides in the twentieth century were the Holocaust during World War II, in which more than six million Jews, gypsies, Poles, communists, homosexuals, physically and mentally disabled persons were exterminated by the Nazis and the 1915 Armenian genocide, in which more than one million Armenians were killed by the Turks. At least 1.7 million Cambodians were the victims of genocide by the Khmer Rouge between 1974 and 1979. More than a million lives were lost and an estimated ten million persons were displaced during the ethnic warfare between Muslims and Hindus in 1947 and 1948 when the former British colony was partitioned between India and Pakistan.

Yet for every noncombatant fatality, there were at least two seriously injured noncombatants or refugees. During World War II and its aftermath, more than 40 million people became refugees; between 2.5 and 4.4 million people were driven from their homes in Bosnia-Herzegovina during the 1990s; more than two million Rwandans were forced from their homeland in 1994. Rape and torture are common fates for those who are not killed in spite of international conventions against these atrocities. During the first four years of the Syrian Civil War, more than 200,000 civilians were killed and there were more than twelve million displaced persons of which more than three million have fled to refugee camps in Turkey, Lebanon, Jordon and Iraq.[4] In terms of human life and suffering, war is the most expensive endeavor that humans engage in.

Defining War

Several definitions of war have been proposed. A common one is a deadly conflict in which there are more than one thousand battle deaths per year,[5] but this definition excludes many deadly conflicts that today are consider to be wars, such as guerrilla wars and terrorist attacks by organized groups such as Al Qaeda, Hamas, Hezbollah, the Islamic Jihad and the Islamic State of Iraq and Syria. It also excludes primitive warfare in such places as the New Guinea highlands or among groups such as the American plains Indians or the Yąnomamö Indians in Venezuela and Brazil, where intertribal warfare consists primarily of raids and ambushes in which there are few casualties in each incident, yet the cumulative causalities over time can amount to a significant fraction of the male population.[6] For our purposes, the definition of war should be broad enough to include primitive war, which includes raids, ambushes, and skirmishes by ad hoc groups, guerilla war, and terrorist attacks in addition to traditional interstate wars. In this book, we will use Margret Mead's definition of war as "socially sanctioned, organized lethal intergroup conflict."[7] which can include guerrilla wars, political and religious terrorism, and ethnic and civil wars as well as interstate wars. However our definition will exclude organized criminal activity such as the Mafia that is subject to a country's internal justice system.

A World Without War?

Many people regard war as a blight on mankind that should be eliminated, and throughout history men have dreamed of this. The 18th century enlightenment philosopher Immanuel Kant looked forward to "a peaceful federation of liberal republics, an enduring and gradually expanding federation likely to prevent war."[8] Although the 1928 Kellogg-Briand Pact, formally known as the Treaty Providing for the Renunciation of War as an Instrument of National Policy, has often been referred to as "an example of the most widely utopian type of scheme, resting on the moral force of world opinion," it was eventually ratified by 62 nations, and it still remains in force. It formed the basis of the legal theory, embraced by Henry Stimson and Franklin Roosevelt at the end of World War II that aggressive war is a crime under international law.

In August 1941, US President Franklin Roosevelt and Winston Churchill promulgated the Atlantic Charter, of which the eighth point laid out their vision of a new world order:

> [A]ll nations of the world, for realistic as well as spiritual reasons, must come to the abandonment of the use of force. Since no future peace can be maintained if land, sea, or air armaments continued to be deployed by nations which threaten, or may threaten, aggression outside their frontiers, they believe pending the establishment of a wider and permanent system of general security, that disarmament of such nations is essential. They will likewise aid and encourage all other practicable measures which will lighten for peace loving peoples, the crushing burden of armaments.[9]

These yearnings for a world without war were incorporated into the United Nations Charter, the preamble of which states, "We, the people of the United Nations, determined to save succeeding generations from the scourge of war, which in our lifetime has brought untold sorrow to mankind . . ." Paragraph 2 of Article 3 of the Charter effectively outlaws war by admonishing "all member states to settle their international disputes by peaceful means in such a manner that international peace, and security, and justice are not endangered" and in Paragraph 3 states, "all members shall refrain in their international relations from the threat or use of force against the territorial integrity or political independence of any state, or in any manner inconsistent with the purposes of the United Nations." However Article 51 of the Charter acknowledges that nations, individually or collectively, have an inherent right to self defense against an armed attack.

Under Article 42 of the Charter, the Security Council may authorize the use of air, sea or land forces of member states to maintain or restore international peace and security. An example is the UN Security Council Resolution 83, passed on June 22, 1950, that recommended that "Members of the United Nations furnish such assistance to the Republic of Korea as may be necessary to repel the armed attack [on the Republic of Korea] and to restore international peace and security in the area." Because the Soviet Union had boycotted the Security Council since January 1950, it was not present to veto the resolution.

Throughout history, men have sought the peaceful settlement of their international disputes. The peace treaty that ended the first war between Athens and Sparta (461–446 BCE) and that was intended to maintain the peace for 30 years included a provision for the arbitration of future disputes between the two citystates, but in the crises that precipitated the Peloponnesian War in 431 BCE, Sparta refused to submit to arbitration because it feared the growing power of Athens.[10] During the medieval period, papal arbitration was occasionally successful in resolving disputes between princes. In 1794, Jay's Treaty between the United States and Great Britain provided for the use of arbitration to settle disputes between the two countries. Until 1971, more than 177 disputes between the United States and Great Britain were settled by arbitration... Article 21 of the Treaty of Guadalupe Hidalgo between the United States and Mexico provided for arbitration of all disputes before either country sought recourse in war.[11]

In response to a proposal by the Russian Tsar Nicholas II, the Convention on the Peaceful Settlement of Disputes was adopted by the Hague Conference in 1899. It established the Permanent Court of Arbitration. Arbitration is not obligatory for the convention's signatories, but between 1899 and 1931, the court was able to decide 30 cases that could not be resolved through diplomatic channels.[12] Although this and other arrangements for the settlement of international disputes have been modestly successful in solving minor disputes, almost no cases serious enough to justify resort to war have been submitted to the Court for Arbitration.

Attempts to Regulate War

Although mankind has not been able to eliminate war, he has sought to regulate it and to limit its tragic consequences. One of the earliest attempts to regulate warfare was that of Augustine of Hippo in the fourth century CE, who formulated standards for going to war—*ius ad bellum*—and for the conduct of war—*ius in bello*. Augustine's just war theory argues that a state may have to resort to the use of force as an instrument of justice for the common good of mankind, but it enjoins rulers against using force unless they are morally certain that their cause is just and that their rights have been violated. There also must be a reasonable prospect that the use of force will produce more good than evil and that it will restore justice.

Attempts to limit the tragic consequences of war in the *ius in bello* tradition are a relatively recent development that began in the 20th century following World War I. The first attempt was the 1925 Protocol for the Prohibition of the Use in War of Asphyxiating, Poisonous and Other Gases, and of Bacteriological Methods of Warfare in response to the German use of poison gas during World War I. This was followed in 1949 by the four Geneva conventions: the Convention for the Amelioration of the Wounded and Sick in Armed Forces of the Field; the Convention for the Amelioration of the Wounded, Sick and Shipwrecked Members of the Armed Forces at Sea; the Geneva Convention Relative to the Treatment of Prisoners of War; and the Convention Relative to the Protection of Civilian Persons in Time of War. More recently, the 1972 Biological Weapons Convention (BWC) and the 1993 Chemical Weapons Convention (CWC) outlaws the development and production of chemical and biological weapons and mandates the destruction of all existing stockpiles of these weapons. The CWC also outlaws the use of chemical weapons as the 1925 convention did for biological weapons. The BWC has been ratified by 154 of the 193 member states of the United Nations, and the CWC by 190 states. By May 2012, the Organization for the Prohibition of Chemical Weapons (OPCW) has eliminated 50,619 tonnes (71 percent) of the chemical weapons deployed in Albania (100 percent eliminated), South Korea (100 percent), India (100 percent), the United States (90 percent), Russia (70 percent), and Libya (85 percent). After Syria's use of chemical weapons against its civilian population in the summer of 2013 was confirmed, the Russian President, Vladimir V. Putin, convinced Bashar al Assad, Syria's president, to have its stockpile of chemical weapons eliminated by the OPCW to avoid a US bombing attack on Syria's chemical weapons facilities. All of Syria's known stockpile of chemical weapons and their precursors were transferred to the OPCW in June 1914 and were destroyed.

The Nuclear Nonproliferation Treaty (NPT), which entered into force in 1970, has been ratified by 190 nations. The treaty recognizes the five permanent members of the UN Security Council—the United States, the United Kingdom, Russia, France, and China—as states that possess nuclear weapons. The other parties to the treaty agree not to develop, test, or deploy nuclear weapons, in exchange for assistance in the development and use of nuclear energy for peaceful purposes. India, Israel, and Pakistan have not signed the treaty, and all three have tested and deployed nuclear weapons. India first tested a nuclear explosive device in 1974 and then again in 1998. Pakistan then responded in 1998 with its own test. Israel initiated its nuclear weapons program in 1949 with the construction of a nuclear reactor for the production of plutonium and began production of nuclear weapons in 1961. It is believed that Israel has a stockpile of about 200 nuclear weapons, but Israel maintains a policy of nuclear ambiguity, neither admitting nor denying that it possesses nuclear weapons. North Korea became a party to the NPT in December 1985 but withdrew from the treaty in 2003. North Korea has conducted three nuclear weapons tests since October 2006. It is believed that North Korea possesses an arsenal of about ten nuclear weapons. Iran has long been suspected of pursuing a nuclear weapons program but agreed in 2014 not to develop nuclear weapons after the UN Security Council and then United States, Russia, China, and the European Union placed increasingly severe economic sanctions on Iran beginning in 2006. In November 2013 Iran entered into negotiations with the five permanent members of the Security Council, the European Union and Germany to eliminate its ability to produce nuclear weapons. Five countries, Libya, Iraq, South Africa, Argentina and Brazil, have given up their nuclear weapons programs.

The 1948 Convention on the Prevention and Punishment of the Crime of Genocide makes genocide a crime that must be punished under international law. It was enacted in response to the Armenian genocide during World War I and to the Nazi Holocaust during World War II. Persons, not countries, can be prosecuted for genocide, and member states are required to prevent and punish acts of genocide. Currently 133 nations have become parties to the convention. The United States Senate ratified the convention in 1988, with the reservation that its citizens could not be prosecuted for genocide without the consent of the US government. Among those countries that have not yet ratified the convention are Chad, Indonesia, and Japan.

Since 1951, there have been six major acts of genocide: during the 1971 Bangladesh war for independence; in Cambodia during the reign of the Khmer Rouge between 1975 and 1979 when approximately 1.7 million Cambodians were killed; in Iraq in 1988 when government forces used chemical weapons against its Kurdish population; in Bosnia between 1992 and 1995 when Serbian forces of the Republika Srpska killed more than 8,000 Bosnian Muslim men and boys; in the 1994 Rwanda genocide, when Hutus killed almost one million Tutsis and moderate Hutus; and in the Darfur region of Sudan, when between 200,000 and 400,000 people were killed, and over two million displaced by Janjaweed mercenaries supported by the Sudanese government.

More than 48 individuals have been tried and convicted of crimes against humanity in various courts and tribunals. They include Slobodan Milosevic, the former president of Serbia, who was arrested and tried before the International Criminal Tribunal for the Former Yugoslavia (ICTY) in the Hague for authorizing the murder of more than 8,000 Muslim men and boys and more than 1,000 women, children, and elderly at Sarajevo in the summer of 1995 and for other crimes against humanity committed by Serbs in the former Yugoslavia. The Milosevic trial lasted from June 2001 until his death in March 2006. General Ratko Mladic, the military leader of Republika Srpska's army in the 1990s, was arrested and brought to trial before the ICTY in the Hague for the 1995 massacre at Sarajevo. The United Nations established the International Criminal Tribunal for Rwanda to try those responsible for the 1994 Rwandan genocide. So far 21 trials have been completed, in which 28 individuals have been convicted of genocide. Jean Kambanda received a life sentence for his role in the 1994 Rwanda genocide. Charles Taylor, the former president of Liberia, was tried and convicted for arming, supporting, and guiding a rebel movement that committed mass atrocities in Sierra Leone during its civil war in the 1990s.

Is War Inevitable?

In spite of all the efforts to eliminate war and to mitigate its tragic consequences described in the previous two sections, human societies continue to fight wars. Since the end of the Cold War, the costs of war in terms of human suffering have increased. Nearly all contemporary wars are between ethnic or religious groups whose objectives are not to inflict a military defeat on their enemy but to eliminate their enemy. Sunni and Shiite Muslims seek to eliminate each other and also others who do not accept their beliefs. Ethnic groups have long fought one another because "they are not like us." The ethnic wars in the Balkans, Darfur, Rwanda, South Sudan, the Caucuses, and other places have been among the most devastating in recent history. When wars were fought mostly between professional armies, they ended when one army lost its ability to fight, usually when it had lost between one-third and one-half of its soldiers or equipment.

Historians, statesmen and political scientists have long sought to find the proximate or immediate causes of specific wars in terms of the political economic, and social factors that brought them about, yet considering the ubiquity of war throughout human history and in prehistoric times, the willingness of men to take existential risks and sacrifice themselves in war when their country, ethnic group or religion is threatened, or to expand the power and influence of their country, ethnic group or religion suggests that war is deeply embedded in human nature and that it has a genetic basis that is embedded in the human genome, and thus is the product of our biological and cultural evolutions. Only ants and a few mammalian species, including chimpanzees, wolves, lions and spotted hyenas, engage in lethal intergroup conflict or war. These considerations raise some fundamental questions: Is warfare part of man's inherent nature, a result of his genetic inheritance and biological evolution, and hence are we doomed to be plagued with it forever? Or is war purely a social phenomenon that is the result of man's social evolution and environment? Or is it a combination of the two? These are the fundamental questions that we will address in the following pages.

Some biologists and anthropologists have maintained that war has been a positive force in human evolution. In the 19th century, Charles Darwin and Herbert Spencer maintained that warfare between primitive tribes promoted human evolution and contributed to the formation of large human societies such as nations and civilizations.[13] However contemporary social Darwinists also believe that "in contemporary societies, war and militarism have become retrogressive and dysgenic, eliminating the 'best elements' of the population on the battlefield."[14]

Chapter 2

The Origin of War

> The Chimpanzee . . . has clearly reached the stage where he stands at the very threshold of human achievement in destruction, cruelty and planned inter-group conflict
>
> Jane Goodall[1]

Humans belong to the taxonomic family of great apes (Homindae) that includes four other species: chimpanzees (*Pan troglodytes*), bonobos (*Pan paniscus*), gorillas (*Gorilla gorilla*), and orangutans (*Pongo pygnaeus*). Figure 2.1 shows the phylogeny (evolutionary history) of the great ape family.[2] Humans and chimpanzees are the only members of the great ape family that engage in lethal intergroup conflict—war. Gorillas live in communities in which a single silverback male presides over a harem of two or more females that live with their offspring more or less independently of other frmales.[3,4] Orangutangs are solitary anmimals that avoid one another except to mate, although they sometimes temporarily congregate in groups.[3] This implies that social living among the great apes originated in the common ancestor of humans and chimpanzees that lived between 6.5 and 7.8 years ago that Richard Wrangham has named *Pan prior*.[5] Cooperative living among the great apes most likely originated with *P. prior* when several gorilla like harems began living together and the females began to cooperate in foraging for food and in raising their offspring. The males provided protection for the group from sexual and other predators in exchange for dependable access to sexual partners.

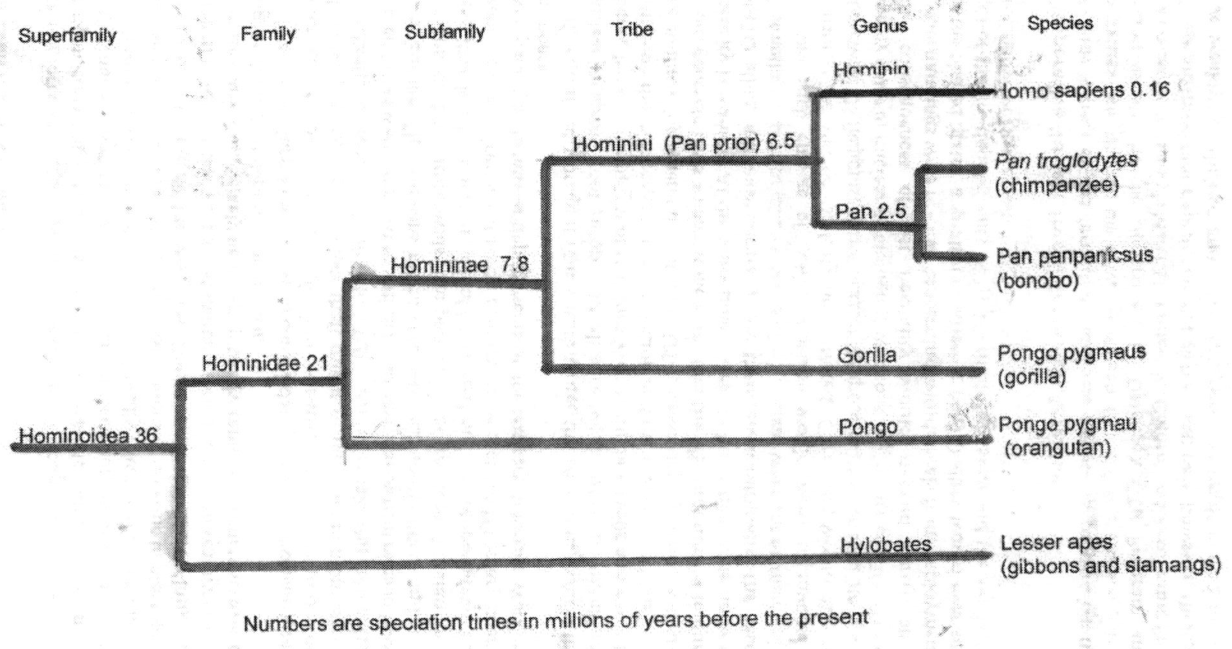

Figure 2.1. The Phylogeny of the Great Ape Family

Because of their cooperative efforts, this group living increased the fecundity of the females. Over time, the *P. rior* founder group slowly grew until it. exceeded the carrying capacity of its environment and then it fissioned into two groups. This growth and fissioning process continued until several of such groups existed. To avoid genetic inbreeding, *P. prior* females migrated into neighboring groups when they reached puberty so that they would not mate with their brothers, fathers, or male cousins, whereas the males remained in their natal groups where they bonded with their male relatives and formed a band of brothers.

Humans and chimpanzees are the only members of the great ape family that engage in lethal intergroup conflict. When two groups of gorillas encounter one another, the silverbacks either ignore each other or express their hostility by displaying, but they have never been observed fighting. Indeed, a group of gorillas usually goes far out of its way to avoid other gorillas.[7] Magdalena Bermajo found that encounters between single male gorillas resulted in either vocal displays (50 percent) or avoidance (50 percent), but not in physical aggression. In 14 encounters between silverback led groups: in nine of the encounters the groups were friendly to the extent of nesting together; in two of the encounters, the groups avoided each other; and two other encounters resulted in a ggressive vocalizations and displays. Only one encounter resulted in physical contact.[8]

Orangutans are solitary animals that avoid each other except to mate and have never been observed fighting other orangutans. This implies that warfare among the great apes originated with *Pan prior*. The alternate hypothesis is that lethal intergroup conflict originated independently in the Pan and Hominid genera, but given the choice between the two hypotheses, the principle of parsimony requires that we choose the simpler of the two hypotheses—that war originated in the common ancestor, *P. prior*.[9]

Group living among *P. priors* led to the development of social bonds among the group members in which they helped and assisted each other. The females cooperated in foraging for food and caring for each other's offspring. The males risked their lives to protect their group's females, their offspring, and the territory in which they live at a cost to their fitness* because they might be killed before their reproductive life spans were over.

---

* Fitness is the number of offspring that an individual is expected to produce during his or her lifetime. It is a function of an individual's environment.

This altruism, however, did not extend beyond their group, a condition known as parochial altruism. These group bonds were so strong that hostility developed toward conspecifics who were not members of the same group, which manifested itself as xenophobia (the intolerance, fear, and contempt that one group of social animals has toward another group of its conspecifics) which is frequently a cause of war. Charles Darwin recognized this when he wrote in *The Descent of Man,*

> In the first place, as the reasoning powers and foresight of the members became improved, each man would soon learn from experience that if he aided his fellow-men, he would commonly receive aid in return. From this low motive he might acquire the habit of aiding his fellows; and the habit of performing benevolent actions certainly strengthens the feeling of sympathy, which gives the first impulse to benevolent actions. Habits, moreover, followed during many generations probably tend to be inherited.[10]

Chimpanzees

Chimpanzees live in a wide belt that extends across equatorial Africa, north of the Congo River, from the west coast of Africa to just west of Lake Victoria and then along the west coast of Africa, north to Guinea. Their habitats range from rain forests and mountain forests on the Ivory Coast to the dry woodlands and mountains of Tanzania and the savanna in Mali.[11] Chimpanzees are mostly diurnal (daytime) terrestrial animals, although some adolescents are arboreal. At bedtime, chimpanzees build their sleeping nests in trees. During the day, chimpanzees feed for six to eight hours and forage over distances of between 1.5 and 15 kilometers. Although they are mostly herbivores, about ten percent of their diet is meat, which includes insects, monkeys, and bush pigs. Because hunting is a cooperative enterprise, food is shared. Chimpanzees may shift their feeding sites seasonally depending upon when and where various fruits ripen.

Chimpanzees live in fusion-fission communities of between 15 and 80 individuals in which the relationship between males and females is loose and flexible, and whose members rarely congregate in one place. The communities are made up of subgroups of between two and twenty or more individuals of all ages and of both sexes that form, breakup and then reform.[12] Individuals of both sexes are free to come and go as they please, and they associate with one another for various lengths of time depending upon their relationship, the reproductive status of the female, and the availability of food. Each animal has its own network of social contacts.

To avoid inbreeding and its genetic consequences which includes, inbreeding depression and reductions in the viability and fertility of offspring, female chimpanzees visit neighboring communities during puberty, and when they become fertile between the ages of 12 and 14 years, they make a commitment and join one community, where they remain for the rest of their lives. In contrast, males remain with their natal communities and form strong affiliative bonds with the other males in their community—a "band of brothers"—that protects their community's females and fight to expand their community's territory.

In most primitive human societies, males remain in their natal societies as chimpanzee males do. In a survey of 865 existing primitive human societies, in 73 percent of those societies, married couples reside in the husband's natal group—a condition called patrilocality—whereas in only 14 percent of the societies, married couples reside in the wife's natal group—a situation known as matrilocality. In 7.5 percent of the societies, married couples can reside with either the husband's or the wife's natal group, and in only 4.2 percent of the societies, married couples reside in neither the bride's nor the groom's natal group.[13] In 20 percent of the non-patrilocal societies, the males become just as closely bonded to one another as they are in the patrilocal societies.

Chimpanzee Territoriality and Border Patrols. Chimpanzees are territorial animals that live in home ranges where they forage for food. The size of their ranges varies between 120 square kilometers in woodlands and up to 560 square kilometers in the savanna. The home range includes a core area, which comprises 35 to 40 percent of the range, and where the community spends between 75 and 90 percent of its time. The ranges of adjacent communities sometimes overlap, so that the communities sometimes come in contact with one another. Male chimpanzees patrol the boundaries of their territory in groups, and the patrols generally consist of three or more males who move quietly along the boundary, looking for evidence of chimpanzees from neighboring communities. They climb trees to reconnoiter the adjacent territory, or stop and sniff the ground and listen for calls of chimpanzees from the neighboring community. Larger patrols may include mothers with offspring looking for food.[14]

Chimpanzees have developed a repertoire of vocal signals that are well understood among them. The most common adult call is the "pant-hoot," or "pant-grunt," which is used in many different situations and can be heard for more than two kilometers even in dense forest. Jane Goodall has catalogued more than 30 variations of the pant-grunt, with meanings of fear, anger, rage, distress, sexual excitement, body contact enjoyment, enjoyment of food, social excitement, and feelings of sociability.[15] Calls are also used to signal arrival at a feeding site, to signal the size of their party, and to assess the size of another group from a safe distance. The calls of a neighboring group can generate both fear and excitement. Upon hearing the calls of a neighbor, a patrol may retreat, hold its ground, or advance toward the rival calls. Although the members of a patrol usually show the same response, females and some males may retreat while the majority advances.[16]

Male chimpanzees are particularly aggressive toward the males of other communities, and some of this aggression is related to territorial defense or to acquiring territory of a neighboring community.[17] When the members of one community come in contact with those of another, the encounter is likely to be hostile, but the nature of the hostility depends on the relative size of the groups, the age and sex of the individuals involved, and if females are present, their reproductive status. Inter-community contacts are most likely to occur at the boundaries or in the overlap areas between adjacent territories, but deep incursions into neighboring territories also occur.

There are two types of border visits: excursions and patrols. Excursions generally involve a large group that includes males, females, adolescents, and infants who forage for seasonal food in a border area. Before moving into a feeding area near a border, these groups usually pause and reconnoiter the feeding area. The males may give pant-hoots and perform charging displays. If there is no response, they will then move into the feeding area.[18] If a patrol encounters a lone individual—male or female—from another community, it may violently attack the individual, beating and biting it until it is severely wounded or killed. At the Gombe Stream National Park—Tanzania's national chimpanzee refuge—on the eastern shore of Lake Kahama, males killed 76 percent of the females that they encountered on patrol, and infants are also usually killed and eaten.[19] If the encounter is with an adolescent female in estrus, a male may groom her and mate with her.

Males are generally fearful of foreign males and usually display hostility toward them. In border patrols, an attack will occur only if the patrol has a significant numerical advantage over the opponent, and usually when the victim is only a single individual. In seventeen encounters cited by Richard Wrangham, eleven involved attacks by parties of three to seven males on a lone male, and in one, the male was accompanied by a single female. In only one of the encounters were the two sides equally matched, both parties consisting of three males and a single female. No incidence of lethal fights between single individuals of different communities has ever been reported.[20] If the parties are equally matched, they will display, exchange pant-hoots, and charge one another, but the outcomes of such encounters are seldom decisive and severe injuries seldom occur, unless a male is isolated from his group during the encounter.

Border attacks by chimpanzees are similar to the raids conducted by primitive human societies, in which a raiding party will lie in wait outside an enemy village and attack only when a single individual can safely be attacked. The decision of whether to attack is based on an assessment of the relative strengths of the raiding party and its victim, and the attack will not occur unless the raiding party has a significant advantage.

Chimpanzees have been observed using weapons against other chimpanzees. They use branches to flail and club their enemies, to flail females who refuse to mate, and occasionally to flail a rival male during copulation. Stones, rocks, and sticks have also been thrown at other chimpanzees, but only about 12 percent of missiles hit their target, so that thrown weapons are usually weapons of intimidation and are seldom used in serious fighting.[22]

Chimpanzee Wars

Chimpanzees have been observed in 34 instances of lethal intergroup conflict between eleven chimpanzee communities at five sites in Africa.[23] Two chimpanzee wars took place in the Gombe Stream National Park in Tanzania between 1970 and 1983. Prior to 1970, a single group o chimpanzees occupied the northern part of the park. Its range extended from Linda Stream in the north to Nyasanga Stream in the south and covered an area of 24 square kilometers, as shown in Figure 2.2(a). The community consisted of 14 mature males, 13 mature females, 7 adolescents, and 5 juveniles that were divided into two subgroups: one that spent more time south of the Kakombe Valley, and the other that

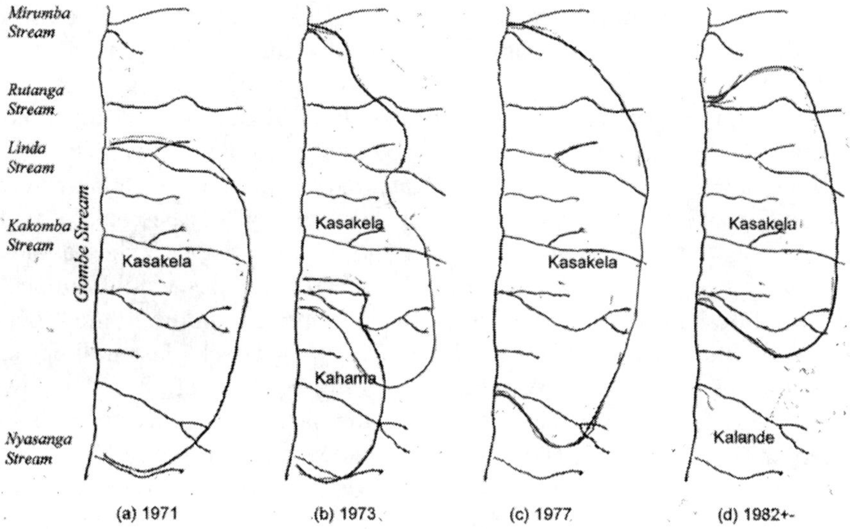

Figure 2.2. Chimpanzee Community Ranges between 1971 and 1982.[21]

spent more of its time to the north of the stream. The northern group included 8 mature males, and the southern group, 6. Three of the 13 females seemed to associate more with the southern males than with the males of the northern group, and later, two other females joined the southern group. Prior to 1971, the two groups occasionally mingled in the Kakombe Valley to feed on an exceptionally fine crop of figs.[25]. Another large community—the Kalande—consisting of at least nine mature males, was located south of Nyasanga Stream.[22]

In 1971 the males of the two groups began traveling in separate patrols, and when the groups encountered each other, the southern group would perform charging displays, which would invariably cause the northern males to scatter. By late 1972, the community had divided into two distinct communities: the Kasakela community in the north and the Kahama community in the south centered on the Kahama stream. The Kasakela community occupied a home range of about 15 square kilometers, and the Kahama community occupied a range to the south of about 10 square kilometers, with the two ranges overlapping about 1 square kilometer as shown in Figure 2.2(b).

In January 1973, the Kasakela males began reconnoitering the Kahama range. A party of Kasakela males traveled to a ridge overlooking the+ overlap area. After they heard calls from Kahama males in the distance, they moved into the overlap area and began to feed. In May, four Kasakela males patrolled the Kahama valley apparently neither seeing nor hearing anything, they returned home. Then a series of killings began in which all of the Kahama males and one female were eliminated over a period of four years. On several occasions, Kasakela male patrols made deep incursions into the Kahama range. W hen a patrol encountered a single Kahama male, the Kasakela males would viciously attack the Kahama male. One Kasakela would pin the victim to the ground while the others hit and bit the victim. The gang attacks lasted about ten minutes and resulted in the mortal wounding of the victims.[27] The two Kahama females may have tried to join the large and powerful Kalande community to the south (not shown), but they were probably killed. The Kasakela males then moved southward and by 1977 had annexed the Kahama range, as shown in Figure 2.2(c). By the end of 1977, all of the Kahama males had been eliminated and the Kasakela began feeding and nesting in the southern part of the range.[23,24]

The second war began when the large and powerful Kalande group to the south of Nyasanga Stream at Gombe, comprising at least nine mature males, began pushing northward and began to move its core area northward. During the 1978–1979 wet season, the Kasakela males began encounting Kalande males north of the Kahama River, who by then had moved into and occupied the former territory of the Kahama community. By 1981, the Kasakela range had been reduced to about 9.6 square kilometers, half of its 1977 size, as shown in Figure 2.2(d). During 1981, the Kalande killed one of the Kasakela males and two of its infants, and by the end of the year, the range had been reduced to six square kilometers, scarcely big enough to support the eighteen adult Kasakela females and their offspring. This forced the Kasakela to move their range further north, so that by the end of 1982 they had enlarged their range to 12 square kilometers. In 1984, a new Kasakela male had matured and five late adolescents joined the Kasakela patrols, thus establishing a new stable balance of power between the Kasakela and the Kalande groups.[25]

War in the Mahale Mountains. A third war took place 200 kilometers to the south of Gombe in the Mahale Mountains, where two chimpanzee communities, M and K, lived. In 1984, the larger M community consisted of 14 mature males and occupied a range of 33 square kilometers, whereas the smaller K community consisted of only five or six mature males and occupied a range of 10.4 square kilometers to the north of the M community's range. Once a year, at the end of the dry season, the M community migrated about five kilometers into the northern part of its home range, where the ranges of the two groups overlapped. Upon the arrival of the M community, the K community relocated northward. After four or five months, the two community returned to their core areas. Then, over a period of six years in the early 1980s, all of the K community's males disappeared, presumably killed by the M community's males, and the M community took over the K community's females. These events may have been motivated by the killing of an M community's infant by the K community's males in 1974 and then by a 1975 retaliatory attack by the M community's males, who killed an infant belonging to the K community.[26]

After that attack, a female of the K community remained with the M community's males for about two hours, during which she was attacked 20 times. Ten months later, the same female with her infant revisited the M community's range and remained there for four days. She was in estrus, and it was obvious that she was attempting to join the M community, but she and her infant were subjected to a brutal gang attack. After all the K community's males had been eliminated, she and the remaining K females transferred into the M community, and all of their infants were killed so the females would come into estrus again.* This resulted in an unusually high ratio of 3.55 adult females to males. Three of the females became pregnant twice, and of the four resulting infants, one was killed and cannibalized by the M males. A fourth immigrant female reappeared with an infant after a six month absence. She was subjected to a brutal attack by three M community males, during which the alpha male seized, killed, and ate the infant. It is highly likely that these infants had all beenl sired by M community males.[27]

Chimpanzee wars, in contrast to border patrols, usually involve deep incursions into a neighbor's territory with the intent of finding and

---

*Infanticide is common by chimpanzee males because while a female is lactating, she will not come into estrus. Chimpanzee females nurse their offsprings for four or five years, during which time, they will not mate and bear infants. The male will kill a female's infant to cause her to become fertile so he can mate with her and she will bear his infant.

attacking a single individual member of the neighboring community.The females of a defeated chimpanzee community are rarely accepted into the victorious community. Captive chimpanzee females may be allowed to survive for a time for the sexual pleasure of the victorious males, but they are usually killed. Male chimpanzees apparently fight to increase the size of their range for the benefit of their own females, whose fecundity is positively correlated with the size of their territory, not to acquire additional females. Some human groups, such as the Yąnomamö, do fight to acquire women, and in these communities, the number of wives that a man has is a measure of his status in his community.

Taï National Park, Four chimpanzee communities inhabit Taï National Park in the Ivory Coast in West Africa. Between 1982 and 1987, the total population varied between 74 and 82 individuals so that the average size of each community was about 20 individuals. Male parties of four to six individuals regularly patrolled the borders of their communities. Unlike at Gombe and in the Mahale Mountains, 87 percent of these patrols involved deep incursions of a kilometer or more into the neighboring community's territory searching for members of the neighboring community. Because of the dense forest at Taï National Park, auditory signals are the primary means for detecting the enemy. If other chimpanzees are heard, the search party begins drumming and screaming. The neighbors usually return the screams, so that the parties can assess each other's size. If females are present, they participate in the drumming to exaggerate the size of their party. The drumming and screaming can continue for 20 to 30 minutes.[28]

If a single individual is encountered, it is attacked, forced to the ground, beaten, and bitten. However no fatalities have been observed in such encounters, because the victim's screaming usually brings help and the attackers leave. If the size of the opposing party is assessed to be larger than the patrol, either the party retreats or reinforcements are called. If an attack occurs, a counterattack has always been observed to follow. If the parties are approximately the same size, they will engage in a meeting engagement.

The Taï chimpanzees fight wars that resemble human set piece battles, and they have developed several battle tactics that include a frontal attack, in which the attackers spread out and attack in a line. If there are many opponents, the charge starts from a greater distance than if the opponent consists of fewer individuals, and the charge is usually accompanied by loud attack calls, or waa-barks. Hitting and biting are the chimpanzees' only weapons, and bleeding bites are frequent. In one encounter, a male was bitten 19 times. Females, both with and without infants, may join a frontal attack or a counterattack, but they usually avoid direct contact with the enemy. They remain behind the line of engagement and join in the waa-barks of the male warriors. In one encounter, five estrous females crossed the battle line to join the males on the opposing side, mated with one or two of them, and then calmly returned to their own side.[29]

The Taï chimpanzees sometimes employ lateral or flanking attacks, in which they try to engage only part of the enemy instead of the entire group, as the Greek hoplites did as described in the next chapter (p. 46). In three such attacks, they were able to engage only a small part of their enemy and chased them for several hundred meters. Such an attack causes the main part of the enemy to retreat and to support their comrades. Flank attacks are usually employed when a small force attacks a larger one.

The third tactic is the raid, which is similar to the raids described above at Gombe. In a raid, the attacking party makes a deep incursion into the enemy's territory, looking for individual victims to attack. The objective of such raids seems to be to scare the victims, rather than to physically harm or kill them.

On two occasions, the commandos found single females with infants that they hit and bit on the head, shoulders, and feet, but they did not inflict serious injuries. Such raids have lasted up to six hours. According to Christopher Boesch and Hedwige Boesch-Achermann, the Taï chimpanzees appear to be able to assess the strength of their enemy and to adapt their tactics accordingly, and they are able to call for reinforcements when needed.[29]

Causes of Chimpanzee Wars. The most likely cause of chimpanzee wars is to expand the size of the community's ranges so as to increase the food resources available to their females. Jennifer Williams and her colleagues have found that at Gombe, the interval between births is inversely correlated with the size of the community's range, i.e., the larger the range, the greater the birth rate.[30] They speculate that chimpanzee males fight to acquire new territory so that their females will reproduce more frequently in the larger territory. Chimpanzee males do not appear to fight to acquire females, as the females of a defeated community who attempt to join the victorious community are seldom accepted but are usually rejected or killed. Instead, the victorious males seek to increase the fecundity of the females of their own community. When the range of the defeated Kasakela community was reduced from 15 square kilometers to 6 in 1981, their 18 adult females lost body weight, their between-birth period increased from four to six years, and the infant mortality rate rose. This evidence confirms the value of large territories.[31]

Bonobos

The bonobo (*Pan paniscus*) is a close relative of the chimpanzee and is sometimes called the pigmy chimpanzee, although the bonobo is only slightly smaller than the chimpanzee. The average weight of a male bonobo is about 49 kilograms, compared to the average weight of a male chimpanzee of 52 kilograms. Bonobos and chimps share a common ancestor that lived about 2.5 million years ago. Bonobos live in communities south of the Congo River in the Democratic Republic of the Congo. In many ways, the behavior patterns of bonobos and chimpanzees are similar. Both live in fusion-fission communities, but bonobos are more gregarious and tolerant of each other and are less aggressive than chimpanzees.

Bonobo communities consist of between 40 and 120 individuals and occupy home ranges of between 22 and 68 square kilometers that may overlap extensively, but communities on adjacent ranges generally avoid each other, because bonobos are much less territorial than chimpanzees. When two parties of bonobos encounter one another, there may be loud vocalizing, antagonistic displays, and occasional fights, but deliberate lethal raids and attacks, characteristic of chimpanzees, do not occur. Bonobo community membership is fluid, mainly because young females who have not yet given birth roam freely among the communities, but bonobo communities are more stable than chimpanzee communities. Subgroups generally consist of between 2 and 15 individuals but may be as large as 40, depending upon the food supply. Subgroups generally form around a female and her immature male offspring, and usually includes some adult males.[32]

Bonobo societies are female dominated. Females break their maternal ties at adolescence and emigrate from their natal communities into new communities in the same way as chimpanzee females do, where they bond into female groups in a manner similar to the way male chimpanzees bond with their male relatives. These female bonded units are the foundation of a bonobo community and its leadership.[33] Female bonobo coalitions dominate males by sometimes attacking and injuring recalcitrant males.[34] Indeed, it seems that female bonding is as strong in bonobo communities as male bonding is in chimpanzee societies.

In contrast to chimpanzee males, bonobo males remain with their mothers throughout their lives and maintain a permanent association with them. The social status of an adult male appears to be about equal to that of his mother, and bonobo males do not attempt to assume a dominant role even among male bonobos. The males are sometimes aggressive toward each other but are rarely aggressive toward females.[35]

A comparison of chimpanzees and bonobos shows that bonobos are sensitive, lively, and nervous, whereas chimpanzees are coarse and hot tempered. Bonobos rarely raise their hair, whereas chimpanzees frequently raise hair to show fear or aggression. Physical violence rarely occurs among bonobos but is common among chimpanzees. Bonobos defend themselves by kicking with their feet, whereas chimpanzees try to pull their attackers close, bite, and frequently kill their opponents. Bonobos are more vocal than chimpanzees, and bonobos copulate face to face as humans do, whereas chimpanzees copulate *more canum,* as dogs and most other mammals do.[36]

Bonobos Make Love, Not War. Bonobos substitute sex for aggression, and their entire social life seems to be centered around sexual activities, so that is impossible to understand their social life without an understanding of their sexual behavior. Unlike most other species, whose sexual life is primarily concerned with reproduction and is a distinct category of their lives, sex has become an integral part of the social life of bonobos, not just between males and females, but also between members of the same sex, between juveniles and adults, and between young adolescents. The frequency of bonobo sexual contacts is higher than that of any other species, and these contacts are strikingly casual.[37] Like humans, but unlike most other animal females, whose estrous cycles are annual and who copulate only during their estrous period, bonobo females engage in sex on almost any occasion during their estrous period, which lasts about 45 days. Because the female estrous cycle is 60 days, they are sexually active 75 percent of the time.[38]

In addition to male-female copulation, both males and females engage in the practice of rubbing their genitals together with members of their own sex, a practice called GG-rubbing, or "koka-koka." Sexual contacts of this type are also common between adults and juveniles, and between juveniles. In one such encounter, an older female's path on a tree branch was blocked by a male juvenile. The female tried to push the male out of her way, but when the male did not budge, she bit his hand, which was clinging to the branch, with no result. Then she rubbed her vulva against his shoulder, which calmed the male, and he moved out of her way. It appeared that the female was close to using force, but instead reassured the young male by means of genital rubbing. In another such encounter, the alpha male of the community regularly chased a younger male into a dry moat as if he were trying to keep the younger male out of the community, but the young male always attempted to return to the group. After a large number of such incidents, the alpha male gave in. He would fondle the younger male's genitals and engage in koka-koka. After that encounter, the young male was allowed to leave the moat, but his first task was to hang around the alpha male to await the cordial signal.[39]

Sexual encounters usually proceed eating. As soon as the caretakers at one zoo approached the bonobos with food, the males immediately developed erections, and before the food was thrown into their enclosure, the animals were inviting each other to have sex. The probable explanation is that whereas food strains relationships in most animal species, for bonobos, anything—not just food—that arouses the interest of more than one bonobo stimulates sexual contact. For bonobos, sex has become a way of mediating conflict. For example, if a cardboard box is presented to two bonobos, they briefly mount each other before playing with the box. Sex keeps competition down at feeding time and facilitates rapprochement in the aftermath of a fight.[40]

Some females when in estrus approach mature males who possess food and, while having sex, takes the food from the male. The females seem to know that sex produces enough tolerance in the males to allow them to remove the food from the male's hands, and they use sex for this purpose. Most of these encounters are rather brief and do not result in ejaculation. Thus, like some human females, female bonobos use sex as an instrument to get what they want.[41]

Sex is also used to promote food sharing, to negotiate favors, to smooth ruffled feathers, and to make up after fights. Such behavior seems to be at the foundation of bonobo society. Females seem as interested in having sex with other females as they are with males. Amy Parish found that female bonobos followed each other around seven times more often than they followed males.[42] Young immigrant females are integrated into a bonobo community through frequent koka-koka with its females, and such activity seems to be the basis of female bonding, that is the basis of the bonobo community.

Bonobo Evolution. As mentioned above, Bonobos diverged from Chimpanzees about 2.5 million years ago and abandoned the chimpanzee aggressiveness and warlike behavior of their ancestors in favor of sex to settle their frustrations and to resolve their disputes. The selective pressure that drove this evolution may h ave been the practice of male chimpanzee infanticide. As also discussed above, female chimpanzees nurse their offspring for four or five years after birth and when they are lactating, they do not come into estrus. Since chimpanzee males' primary interest in females is in procreation, males frequently kill a female's infant, which causes her to stop lactating and to begin her estrous cycle again. Because the males usually cannot identify their offspring from others in the community, infanticide usually occurs in groups that a male is attempting to take over, or of the offspring of females who are strangers to them.[41]

Since the pleasures and soothing effects that sexual intercourse has on both male and female bonobos, the nonreproductive sexual practices spread through the entire bonobo population and same sex sexual practices became fixed in the population. Because the bonobo birth rates are about 20 percent greater than that of chimpanzees.[43] These practices could have become fixed in the bonobo population by natural selection.

Chapter 3

Prehistoric Warfare

All civilizations owe their origins to the warrior.

John Keegan[1]

Human warfare can be divided into two main epochs, primitive warfare and modern warfare, which are separated by the agricultural revolution and the rise of civilizations that began about 10,000 years ago in Southeast Asia. Before the agricultural revolution, humans lived in small hunter-gather foraging bands of no more than 150 individuals and in most cases fewer than 50. Since there are no written record of human history before about 5,000 years ago, we don't know when *Homo sapiens* or their ancestors began to fight wars, so we must rely on the archeological record and on anthropology for our knowledge of prehistoric warfare. Archeology provides physical evidence of prehistoric warfare prior to the invention of writing in the form of human skeletal remains, cave art, sculpture, painted pottery, weapons, and fortifications that primitive warriors have left behind. There are, however, missing pieces in the archeological record because much of the physical record, such as wooden spears, bows, and arrow shafts have disintegrated over time and there is no record of the motives for warfare in prehistoric times.

Another source of how primitive societies lived and fought are accounts of currently existing primitive societies by anthropologists and other observers, but these contacts began only two or three centuries ago. Unfortunately, there are very few primitive peoples living today who have not fallen under influence of modern civilizations, so that currently existing primitive societies may not be representative of prehistoric ones. Furthermore, many of the accounts of explorers and traders who were the first to encounter primitive societies are biased by the cultural prejudices of the observer. Some primitive societies that are known to have engaged in lethal killing in the past, such as the !Kung bushmen of southern Africa, have been pacified by European colonialists and missionaries and are now regarded as peaceful societies. Many anthropologists who had investigated these societies in the late nineteenth and early twentieth centuries were either not aware of, or neglected these pacification efforts and have created the myth of the "peaceful noble savage," that may have never existed.[2]

Four ethnographic surveys of contemporary primitive societies have shown that warfare has been ubiquitous among them. In one cross-cultural survey of 90 societies by Mark Ross, only 12 were found to "rarely or never" engage in warfare[3] but six of these were tribal or ethnic minorities that had long been subjects of the peaceful administrations of a modern colonial power, three were agricultural tribes that lived in geographically isolated locations, and the other three were nomadic hunter-gatherer tribes living in either equatorial jungles or the arctic tundra, where population densities are low. In another study of 50 societies by Keith Otterbein, only five "infrequently or never" engaged in warfare.[4] Four of these had been forcefully driven into isolated refuges, where they were protected from being attacked. In a third study of 157 North American Indian tribes and bands, only seven said that they had never engaged in raiding or in any type of warfare.[5] All seven were small nomadic bands residing in the driest, most isolated regions of the Columbia Plateau and the Great Basin of North America.

Evidence of Prehistoric Warfare.

The earliest evidence we have of warfare among hominids comes from the analysis of 36 Australopithecus* skeletons from South Africa that lived about 2.5 million years ago, 19 of which show depressed skull fractures and one of which has a stone embedded in its skull. Such injuries most likely were caused by other Australopithecus individuals.[6] Fossil remains of six *Homo antecessor*, another extinct hominid species that lived between 1.5 million and 800,000 years ago, from northern Spain show cut marks that indicate that "the flesh of at least six people of all ages from Upper Dolina were put to a very calculated, meticulous and

---

* Australopithecus was an extinct hominid genus that lived between 4 and 2 million years ago. in southern Africa.

Figure 3.1 – Paleolithic Cave Art of Fighting Archers from Southern Spain.[7]

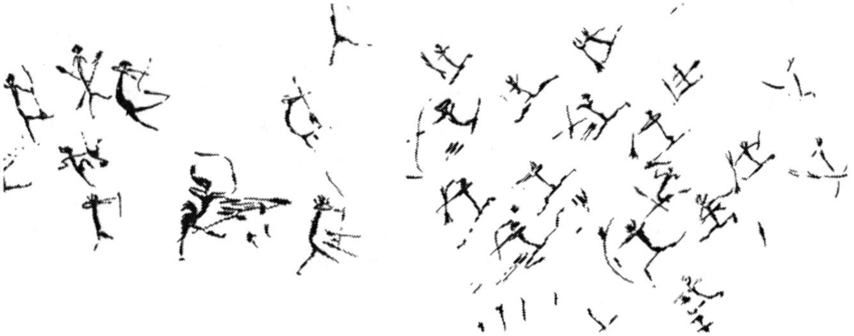

Figure 3.2 – Battle Scene From The Gasulla Gorge in Southern Spain.[8]

efficient use, as if they had been game animals captured for meat."[9]

Evidence of violent homicide among humans has also been found at several locations in Europe that date from the Upper Paleolithic period between 36,000 and 24,000 years ago. In a cemetery in Czechoslovakia dating from the time that *H. sapiens* first arrived in Europe, a mass grave containing the skeletons of men with fractured skulls together with women and children implies that violent conflict among humans occurred at that time. At Grimaldi in Italy, a child's skeleton has been found with an arrowhead embedded in its spine that dates from between 27,000 and 36,000 years ago, and a skull showing cut marks indicating that the individual may have been scalped was found in Southern France dating from the same period.[10] Although these deaths were the result of human violence, we don't know the circumstances under which they occurred, and they may not have been the result of warfare. Just south of the Egyptian-Sudanese border at Gebel Shaba, Cemetery 117, dating from between 12,000 and 14,000 years ago, contains 59 skeletons of humans who were evidently killed by weapons. The cemetery contains the skeletons of 20 adult males, 21 adult females, 11 children, and seven adults whose sex could not be determined. Forty percent of these skeletons contain small stone-flake points that are believed to be spear points or arrowheads. Some of the skeletons suffered multiple wounds. One adult female skeleton contains 21 chipped stone artifacts, and one adult male skeleton contains 19 artifacts. All of the children suffered wounds in the head and neck. Some of the adult skeletons show healed fractures of their forearm bones, a common trauma of victims of violence.The circumstances of these traumatic deaths cannot be determined, but it it is likely that such a large group were casualities of war.[11,12]

The earliest man-made records of warfare are cave paintings found in Western Europe depicting warfare dating from the Upper Paleolithic period. Figures 3.1 and 3.2 are cave paintings of battle scenes from the Morella la Vella and the Les Dogues shelters in the Gasulla Gorge in Southern Spain that date from between 12,000 and 10,000 years ago, about the time the bow and arrow are thought to have been invented. In Figure 3.1, the warriors on the left appear to be directing flanking fire against their opponents on the right. Figure 3.2 depicts a battle between archers from the same period. These paintings appear to indicate that warfare circa 12,000 to 10,000 years ago had developed to the point of the use of organization and tactics.

Hunter-Gatherers and Territoriality

For many years, it was commonly believed that primitive hunter-gatherers were peripatetic, foraging over large areas, and did not have permanent residences. They did not claim an exclusive right to a particular piece of territory where they lived and defended it against intruders. Ethnologists later discovered that some primitive hunter-gatherer groups are territorial because they claim and defend territories in which they permanently reside and foraged and hunted for their food and other needs. Keith Otterbein found that 64 percent of the primitive societies in his survey of 50 fought to defend their territories.[13]

Territorial foragers find an area that is better suited than others in providing for their needs, and they settle there and take exclusive possession of it, defending it against intruders. A territory usually consists of a core area where the group resides and of a larger range where they forage and hunt for food and other needs. The core area may provide shelter in the form of caves or cliffs, or primitive shelters may be built for protection from the weather.

Whether to settle down and become territorial depends on the trade-off between the costs and the benefits derived from being territorial. The costs include the time and energy required and risks involved in defending the territory. Larger territories are more costly to defend, and the size of the territory depends on both group size and the density and dependability of the resources within the territory. In harsh climates such as arctic and desert climates, resource densities are low and territories tend to be large to accommodate groups whose size may range between 25 and 50 individuals. In temperate and tropic climates, some locations are better suited than others because they are richer in food resources or are more defensible. Locations along riverbanks, in river deltas, and along ocean coasts may be more desirable than other locations because they provide fish and water fowl and the ground tends to be moist and to have a greater abundance of fruits and berries. In deserts, having access to a well or water hole is vital for survival.[14,15]

The lands occupied by the Western Shoshoni Indians in Eastern Idaho, Western Wyoming, and Northern Utah were so arid that the population density was only one person per fifty square kilometers, so that the occupants were forced into a nomadic existence and families shared information on the location of good food sources. They seldom congregated in large groups and did not live in villages. They had no concept of land ownership.[16]

There are two kinds of territorial defense: perimeter defense and social boundary defense. In perimeter defense, the perimeter of the territory is marked and patrolled, and trespassers are challenged and fought. Groups that employ perimeter defense usually regard trespassing as a declaration of war. In some places, territorial groups coexist with nomadic bands that frequently invade and raid their territories, which the resident tribe then defends to the extent of killing the invaders. The Ona Indian tribes of Tierra del Fuego in South America raided their neighbors' territories whenever the neighbors trespassed on their hunting grounds in pursuit of guanaco. The Washo

Indians who lived in California and Nevada attacked nomadic bands when they fished in lakes and hunted deer in their core areas in the valleys of the Sierra Nevada Mountains. The Nyae Nyae bushmen in Namibia, Africa, kill neighbors who invade their territory to gather vital plant foods. Territorial invasion has been one of the principle causes of war among primitive hunter-gatherers.

Social boundary defense is frequent in places where resources are scarce or unpredictable such as deserts and arctic regions. In this type of defense, outsiders may be granted permission to the resources of the territory depending on the availability of a variable resource such as a water hole. In the Australian desert, the Walbiri Aborigines would allow other bands to use their water holes only with their permission.[17] The Paiute Indians in the Owens Valley in California occupied fertile areas with dense stands of pinion pines and abundant game along the Owens River in the Sierra Nevada mountain range. The Paiute were organized into bands, each comprising several villages. They divided the valley into territories with marked boundaries that crossed the river and extended into the Sierra Nevada Mountains on both sides of the valley. These boundaries were enforced by social and religious sanctions, which were reinforced by occasional threats of war if the boundaries were violated.[18]

The benefits and costs of territoriality can be assessed by comparing the fitness of territorial females with those living a nomadic existence, or by comparing the amount of energy expended or gained by populations in the two environments. Energy may be saved by easy access to resources and dependable shelter, but lost in the increased energy expended in defending the territory.

Warfare among Hunter-Gatherers

Australian Aborigines. At the beginning of the nineteenth century, the Australian aborigines were among the most primitive peoples on earth. They had arrived in Australia about 50,000 years ago, and remained undisturbed until the British arrived in Australia at the end of he eighteenth century. The aborigines consisted of about 300,000 individuals living in between 400 to 700 bands that averaged between 500 and 700 individuals. Although the abroigines were hunter-gatherers, they were also territorial.

The British found that the abroigine bands fought one another and possessed weapons that were used exclusively for war, including clubs and spears, the atlatl (sear thrower) and a hooked boomerang with a sharp pointed end. Although they seemed to have had little reason to fight according to modern standards, warfare was endemic among them. The principle causes of fighting were revenge for previous perceived wrong s against the band or one of its members, continuing feuds, and the defense of their territories against other forger groups, for which alliances were often formed with other groups. No man's lands were established between their territories, where they sometimes traded.[19]

The earliest description of warfare among the Australian Aborigines was written by William Buckley, an English boy, who at the age of 13 was taken to Australia in 1803 on the first penal ship to arrive there. Shortly thereafter, he escaped from the penal colony and lived among the Aborigines for 32 years. Buckley describes Aboriginal warfare as consisting of night attacks, massacres, and treachery. Women sometimes joined in the fighting. He describes one episode:

> The next day we moved on to another fresh water lake of considerable extent, where we camped, not very much at our ease, as we saw another tribe on the opposite shore. In the middle of the night we heard a dreadful uproar in that direction, and in the morning learned that those we had seen before dark had been fallen upon by some others whilst they were sleeping; so upon hearing this we went to their assistance. On our arrival a horrid scene presented itself, many women and children laying about in all directions, wounded and sadly mutilated. Several of the poor creatures had rushed into the lake and drowned.[20]

Of another such occasion, Buckley wrote,

> [W]e were unexpectedly intruded upon by a very numerous tribe, about three hundred. Their appearance coming across the plain, occasioned great alarm. . . .On the hostile tribe coming near, I saw they were all men. . . . In a very short time the fight began. . . . Men and women were fighting furiously, and indiscriminately, covered with blood. Two of the later were killed in this affair. [The battle ended with at least three killed, and a counter attack was planned for that night.] Finding most of them

> asleep laying about in groups, our party rushed upon them, killing three on the spot and wounding several others. The enemy fled . . . leaving their war implements in the hands of their assailants and their wounded to be beaten to death by boomerangs.[21]

On Australia's northern coast, the Murngin Aborigines also engaged in extensive warfare. In the early 20th century, ethnographer Lloyd Warner estimated that during a 20-year period in the late 19th century, of an overall population of about 3,000, about 800 were fighting-age males, and of these, 200 had been killed in battles. According to Warner, "Warfare is one of the most important social activities of the Murngin peoples and surrounding tribes," and most of the fighting was to avenge the death of relatives, the stealing of women, accusations of death by sorcery, and sacrilege.[22] This fatality rate of 25 percent is comparable to the fatality rates among New Guinea farmers who also engaged in intense warfare. The Murngin had names for six different types of warfare, including surrounding a camp at night, a pitched battle, and a general open fight.[23]

North American Indians

Plains Indians. The North American Plains Indians occupied the Great Plains between the Mississippi River and the Rocky Mountains, and from Texas and New Mexico to the southern parts of the Canadian provinces of Manitoba and Saskatchewan. In 1800 there were between 100,000 and 150,000 Plains Indians organized into 23 tribes, or "nations," whose populations ranged between 300 Kiowa Apache to 27,000 Sioux and 30,000 Blackfoot.

At the beginning of the eighteenth century, the Plains Indians had adopted two different lifestyles. Those like the Crow were nomadic hunter-gatherers who followed the great buffalo herds that inhabited the plains, and those like the Caddoans and Navaho Apache who led seminomadic lifestyle according to the seasons. From spring until harvest time in August, the seminomadic tribes lived in villages along the rivers of the plains, where they practiced horticulture, growing maize, watermelons, pumpkins, wheat, and kidney beans, but in the fall and winter, they reverted to a nomadic life of hunting buffalo, which was the major source of meat and skins for the Plains Indians. Before the Indians acquired horses from the Spanish in the 17th and 18th centuries, they followed the buffalo herds on foot , killing them by driving them into *pistins* (traps) or over cliffs.[24]

Warfare and raiding were an important part of the life of an Indian male, and for the Plains Indians, being a skilled and competent warrior was considered a necessity. Unless a male was recognized as a competent warrior, he could neither marry nor gain the respect of his fellow tribesmen. Indians went to war for three reasons: to protect themselves and their territories from intruders; to revenge previous raids, ambushes, and perceived wrongs; and to demonstrate their bravery and their abilities as warriors.[25] Young boys were trained to be warriors. A boy received his first bow and arrows as young as the age of four from his father, who taught him the rudiments of shooting and hunting small game. When a boy was large enough to straddle a horse, he received a colt and was taught to ride. At the age of ten, he learned to throw a lance, dodge arrows, and take advantage of cover. When he was 15, a boy might be invited to join a warrior society, and by 17, he would have stolen a horse, taken a scalp, or *counted coup*—touched a brave of another tribe with a *coup* stick held in his hand. A man could not marry until he had earned the right to, which, for the Crow Indians, meant that he either had reached the age of 25 or had *counted coup*.[26]

The Plains Indians engaged in three kinds of intertribal hostilities: raids, ambushes, and battles. The most common form of hostility was the raid in which a raiding party attempted to *count coup* against the braves of another tribe and, after the Plains Indians acquired horses from the Spanish, to steal horses from another tribe. Raiding was as much of a sport as it was a form of serious warfare. It provided an opportunity for a young brave to demonstrate his courage and to achieve honor and respect as a warrior. A raiding party usually consisted of between four and twenty warriors. Raids were organized by a warrior who would invite men with proven or potential ability to accompany him. The invitation could be accepted or refused without any stigma. The raid might also include several younger uninvited braves who were eager to become warriors.[27] Because the tribes were widely separated geographically, a raid would generally last between several weeks and several months. On foot, a party could travel about 25 miles a day, but after Plains Indians had acquired horses, a party could travel at an average speed of 50 miles a day. The raided tribe would generally pursue the raiders, so that unless the raiders got a good head start, a skirmish between the two parties was likely to occur, which could result in fatalities.[28]

A war party could be as large as several hundred men, all of whom were volunteers. After the Indians had acquired horses, there was little opportunity for a surprise attack, so battles became frequent and took place on the open plain. Once a battle began, there was no organization or formation and it was every man for himself. The battlefield was hence one of chaos; the din was incredible; horses ran into each other; clubs, tomahawks, and lances were swung in every direction; and the dust was so thick that one could hardly see.

Frequently, a man would mistakenly kill one of his companions. Casualties were generally low, however. If a single Comanche fell, his comrades would disengage even if the odds were overwhelmingly in their favor. The mutilation of fallen enemies, either by scalping or taking of whole body parts, was a common practice. Many tribes believed that scalping killed the soul, and a stretched and dried scalp provided the warrior with a prop to tell the story of its taking. Sometimes a fallen body of an enemy was taken home for the women and children to *count coup* on. A common practice was to collect finger bones for necklaces. Torturing prisoners was common among the Comanche and Apaches, but other tribes, including the Sioux, Cheyenne, and Arapahos, did not torture their prisoners. Women and children prisoners were frequently adopted into the captor's tribe, but in a revenge raid, they were as likely as an enemy warrior to be killed.[29]

Eastern North American Indians. The heavily wooded topology and numerous streams and lakes of the eastern part of North America influenced the nature of Indian life and warfare in that region. The eastern Indians were as warlike as those on the plains, and they fought for many of the same reasons: intertribal feuds, gaining of prestige and power, demonstration of their courage and martial skills, avenging of real and perceived wrongs, acquisition of hunting territory and fishing rights, extracting of tribute, and domination of weaker neighbors. Like the Plains Indian, raids and ambushes were the prefer-red forms of combat, but the woodland Indians also engaged in extended sieges of fortified villages, and occasionally in battles involving large numbers of warriors.[30] The tribes below about the 45th parallel of latitude both hunted and engaged in horticulture. They lived in villages, which they sometimes fortified with palisades of upright logs as shown in Figure 3.3. The palisades provided an effective defense because few braves were willing to endure the hail of arrows to climb over the palisade. Instead, the attackers chose the safer alternative of a long-term siege.

Figure 3.3. The Fortified Coastal Algonquian Village of Pomeiooc (an engraving from a 1585 painting by John White, a Roanoke colonist)

Extended sieges were seldom successful because the defenders would lay up an adequate supply of food and there would be a source of water within the palisade. Occasionally, an open battle outside the palisade would take place. Usually, as on the plains, casualties were low.

The Eskimos. Because the Eskimos of Alaska and northern Canada lived in the harsh arctic climate in widely separated settlements, it might be thought that they were too preoccupied with their survival to fight among themselves, but prior to their pacification by the US Army and the Royal Canadian Mounted Police in the 1870s, the Eskimos engaged in extensive warfare.

Based on interviews with older Eskimo men in 1969 and 1970, anthropologist Ernest Burch described warfare among the Eskimos of Northwest Alaska in the 19th century. Between 10,000 and 12,000 Eskimos lived in Northwest Alaska in the mid-19th century. They were organized into approximately 21 tribes whose size varied between 200 and 1,200 members. Intertribal relations included intermarriage, the migration of both individuals and whole families between tribes, feasting, and highly structured trading, as well as feuding and warfare.[31]

In spite of their friendly relations, warfare was continuous and ubiquitous among the Eskimo tribes, and in addition to fighting among themselves, they also fought with the Southwest Alaskan Eskimos and the Athapaskan Indians in the Alaskan interior. The Eskimo male's attitude toward war was similar to that of the Plains Indian male: A male's reputation within his tribe was based on both his mental and physical toughness and endurance. Males competed with one another both within the tribes and with males of other tribes in tests of physical strength and agility, dancing ability, storytelling, and cleverness in trading. Disputes arising from these ostensibly peaceful competitions between members of different tribes often escalated into armed conflict. The principle cause of warfare was to avenge grudges of individuals toward members of other tribes, which were remembered for years, and when a sufficient number of these grudges had accumulated, a war usually resulted. At least one member of every tribe had unresolved grudges against members of another tribe, so that every tribe had reasons to go to war,[32] but because the tribes also maintained peaceful relations, a serious reason was needed to precipitate hostilities, and murder was often the reasom.

The decision to go to war depended on two factors: the seriousness of the grievance held by the offended individual in relation to the overall relationship between the two tribes, and the assessment of the likelihood that the war party could win the engagement. A tribe would not send out a war party unless it felt that that its chance for success was very high. The war decision was usually made by the aggrieved person and did not depend upon the consensus of the tribe. Nevertheless the aggrieved individual usually consulted with other members of the tribe, but it was up to that individual to recruit a war party from among the other males of his tribe, usually his male relatives. The objective was usually to kill a specific individual or individuals in the enemy tribe.

Participation in a war party was voluntary, as it was with the Plains Indians. It was important that the war party be large enough to defeat the enemy, and the final decision to attack was often made just before the attack, when the strength of the enemy could be firmly determined. If the village contained more houses than expected, implying a greater strength of the enemy, the attack might be called off. Similarly, when two opposing forces were about to engage one another in an open battle, if one side was significantly weaker than the other, the weaker side was likely to attempt to withdraw before the battle, or it might attempt to form an alliance with another tribe and enlist its aid. Sometimes the alliances were made during the summer rendezvous among the tribes, and they were usually between neighboring tribes. Such alliances might not last for long, and tribes in an alliance one year may be at war the next.[33] Among the Eskimos, the most common form of warfare was raiding, as it was among most primitive societies, but open battles also occurred. A raiding party would typically consist of between 15 and 20 men, although it could number as many as 50. Raiding parties always had a leader, and the better disciplined unit usually won.[34] Anthropologist Edward W. Nelson, an early authority on Eskimo warfare, has described a raid as follows: "[T]he old way of making war among the Bering Eskimos was to lie in wait around a village until night, then steal to every house and barricade the doors from the outside. The men of the village being thus confined, the attackers could leisurely shoot them with arrows through the smoke holes [in the roof.]"[35]

Eskimo villages were usually located to minimize the likelihood of surprise attack: on a spit of land, or adjacent to a thicket of willows so they could be easily defended. Tunnels were dug between dwellings, and dogs were used as sentinels. Sometimes sharpened spikes of caribou bones were placed in the ground in large clusters that would easily penetrate an Eskimo moccasin. The defenders would charge the attacking force and drive it into the cluster of stakes. Because the Eskimos used dogs as sentinels, surprise was difficult to achieve.

The objective of Eskimo warfare was to kill, and women and children were not spared. Taking booty was never an objective of Eskimo warfare. Children were killed because the Eskimos believed that, if left alive, children would avenge the deaths of their relatives when they grew up. Burning logs and bark were thrown into houses to set them on fire and to force the inhabitants to vacate so they could more easily be killed. An attacked village seldom surrendered, because the Eskimos seldom took prisoners, and to surrender meant death. If the battle ended in a draw, the two forces might agree to meet the following day outside the village and resume the fight.[36]

An open battle took the form of two lines facing one another in an open field in broad daylight. The preferred weapon was the bone bow and arrow, although clubs were also used, and armor in the form of plates of bone and ivory was worn like coats of mail. The battle would begin with a prolonged firefight. Each side would form a line facing the opponent, but not in ranks. The objective of the firefight was to eliminate the opponent, but failing that, each side attempted to goad the enemy into firing all of its arrows with the minimum amount of damage by dodging the enemy's fire.[37]

Yąnomamö Indians

The Yąnomamö Indians of South America number between 15,000 and 22,000 individuals. They live in about 200 politically independent villages of between 50 and 400 people at the headwaters of the Orinoco River where the borders of Venezuela, Brazil and Columbia converge. They are subsistence horticulturists who have not domesticated animals, except dogs, and they obtain their animal protein from hunting. Except for the steel machetes and metal pots that were given to them by missionaries in the 1940s and 1950s, the Yąnomamö still use stone and bone for their tools and weapons and remain until this day Neolithic horticulturists. Although between 80 and 90 percent of their food comes from their gardens, the Yąnomamö spend as much or more time hunting as they do farming. The most common game are large birds similar to our turkey and pheasant, wild pig, armadillo, alligator, deer, and many varieties of fish, insects, larvae, and crabs. They also gather palm fruits and honey that is a prized delicacy.[38]

The relationships between Yąnomamö villages are complex and range from being trading and feasting partners who exchange women for marriage to being hostile and raiding one another. According to Napoleon Chagnon, who has spent many years of study and in fieldwork among Yąnomamö, the primary objective of the Yąnomamö is "to preserve the sovereignty and independence of their villages in a milieu of chronic warfare. . . . To maximize their chances for independent political existence in this milieu, members of sovereign villages protect their autonomy by adopting an agonistic stance towards neighboring groups."[39]

To maintain their independence, Yąnomamö villages form alliances with one another, which tend to be fragile and can dissolve rapidly. Alliances between villages are usually the result of a sequence of feasting, trading, and ultimately the exchange of women for marriage. A Yąnomamö village cannot simply ask another village to be its military ally. To do so would be to admit its weakness and vulnerability. Instead, it will either invite the potential ally to a feast and attempt to develop a trading relationship with it, or it will visit the potential ally, bringing goods for trading and food for feasting. An alliance relationship is developed over a period of months or sometime years.

The exchange of women is the most intimate form of an alliance, but an alliance may materialize at the trading and feasting stage. However alliances that do not pass beyond this stage are weak alliances, but they do serve to limit the hostilities between their members, unless provoked by a specific incident such as the abduction of a woman.[40]

Conflict among the Yąnomamö. The Yąnomamö have a reputation for violence and have been called the "fierce people." They have a code of ferocity—*waiteri*—by which men are obliged to demonstrate their aggressiveness and bravery. Boys are taught to be violent at an early age and grow up to be violent toward women. Normally, villages that feast together regularly do not fight one another; however even the best of allies will occasionally agree to end a feast with a chest-pounding duel to demonstrate that even though they are friends, they are willing to fight to demonstrate their *waiteri.* Any feast can end in violence because of a long remembered incident in the past. Feasts and alliances often fail to establish a stable, amicable relationship between two villages. When this happens, the villages may coexist for some time without any overt expressions of hostility, but this is a metastable situation. They must become allies, or they will eventually develop a hostile relationship.[41]

The Yąnomamö have a graduated scale of conflict ranging from chest-pounding and side-slapping duels to club fights and village raids at the upper end of the spectrum of violence. Chest-pounding and side-slapping duels are governed by stringent rules and hence are analogous to nonlethal sports in other societies. Chest-pounding duels are generally precipitated by such minor affronts as malicious gossip, accusations of cowardliness, stinginess with food, and niggardliness in trading.

Duels are usually between members of different villages and take place after a feast. One chest-pounding duel observed by Napoleon Chagnon was precipitated by a visiting village that had abused its invitation to a feast. The village had arrived a week before the feast and demanded to be fed. After the visitors had made gluttons of themselves and had raided the host village's garden, the hosts then challenged their guests to a chest-pounding duel. A man from each village stepped forward. One spread his legs, and he extended his chest toward his opponent with his hands behind his back. His opponent then adjusted the man's chest and arms to give himself the greatest advantage and then stepped back to deliver his blow. He wound up like a baseball pitcher but kept both feet on the ground and then delivered a blow with all his might to the victim's left pectoral muscle. The victim's knees buckled and he staggered for a few moments and shook his head to clear the stars but remained silent. Such a blow would invariably raise a painful lump on the victim's chest. The victim might receive as many as four blows and then demand his right to strike his opponent with as many blows as he had received. If the fighter who had delivered the first blows was injured before receiving the number of blows he had delivered, he could retire and be replaced by a comrade, who could then deliver more blows to the second fighter than the first fighter had delivered. Thus, good fighters were at a disadvantage because they would receive more blows than they had delivered. Their reward was that they earned the reputation of being a *waiteri*—a fierce one.[42]

The next level of violent sport is the side-slapping duel, which is similar to the chest-pounding duel, except that the blows are delivered with the side of the hand to the area between the rib cage and the pelvis. Such blows can produce more severe injuries than chest-pounding. A side-slapping duel is sometimes the result of the escalation from a chest-pounding duel. In one side-slapping duel reported by Chagnon, when an influential man of the host village was knocked unconscious, his compatriots took up their bows and arrows, whose tips were coated with poison, drew them, and pointed them at their opponents. The visitors then picked up firebrands and, holding them as weapons, withdrew from the village.[43]

Club fights are the third level of violent games. Most club fights result from disputes over women and food theft, and they can take place both between members of the same village and between villages The wooden clubs are between eight and ten feet long, are quite heavy, and resemble large pool cues. They are sharpened at one end so they can also be used as spears. The club is held at its thin end. An enraged husband may challenge a man who has been observed or suspected of trysting with his wife to a club fight. The suspected culprit first strikes the challenger on his head with his club. After he receives the blow on his head, the challenger can then deliver a blow to his opponent's head. In many fights when blood starts to flow, everyone takes a pole and joins in the fighting. The tops of the heads of many of the older men are covered with ugly deep scars of which they are immensely proud, and they shave the tops of their heads and rub red ocher on the bare skin to make the scares more prominent. Club fighting is more prevalent in larger villages because there are more opportunities for illicit sexual liaisons.

Many villages break up or fission because of club fights over women. In one incident, a man took the wife of another man, claiming that she had been mistreated by her husband. A club fight ensued in which nearly every man in the village participated. The fighting escalated when the husband jabbed his rival with the sharp end of his pole and badly wounded the rival. The village headman tried to keep the fighting restricted to clubs, but when the husband's rival speared the husband, the headman went into a rage and speared the rival, running his spear through the rival's body, killing him. The wife was then returned to her husband, who punished her by cutting off both her ears with his machete. The kinsmen of the dead rival were then ordered to leave the village to avoid further bloodshed. They joined two other villages that were at war with the first village because they would then have an opportunity to raid their native village. Because of its high fatality rate, club fighting is infrequent among the Yanomamö.[44]

The forms of conflict described above are more suggestive of violent sport than of war because they are governed by rules intended to avoid fatalities, although the injury rate appears to be high compared to the injury rates of what is normally consider to be sport. Village raiding, on the other hand, fits our definition of war as "socially sanctioned, organized lethal intergroup conflict." Yąnomamö wars consist of sequences of village raids that are motivated by a need to revenge a previous wrong such as a previous killing, raid, wife theft, suspicion of sorcery, or a club fight over women. The Yąnomamö themselves regard fights over women as the primary cause of their wars.[45]

A raiding party usually consists of between eight and twelve men. They time the departure from their own village so they will arrive at the village to be raided before dawn. The raiders approach the village during the night and lay in ambush along a path leading to a latrine or a water source. When a single person from the enemy village appears, she or he is attacked. Men and boys are killed, but women and girls are often taken captives. If the village suspects a raid, it will take defensive precautions. Only large, well-armed groups will leave the village, because the raiders will not attack such a group.[46] The primary purpose of most raids is to kill one or more of the enemy and then to escape without being challenged. If one of the raiders is killed, then the raid is considered a failure, no matter how many of the raided villagers are killed. Few raids are initiated with the objective of abducting women, because captured women are regarded as a side benefit of a raid. A captured woman is usually raped by all the men in the raiding party, and later by the other men of the village who had not participated in the raid, and then is given to one of the men as a wife.[47]

Primitive War in Papua New Guinea

The Papuan people of the New Guinea highlands live in villages of about 120 individuals. A village consists of about 15 houses, in which between five and forty people live. The males sleep together in a dormitory in the front part of the house, whereas married women occupy separate quarters with their children, in either a partitioned area in the main house or in single room huts. The owner of the house is the titular head of the household who enforces discipline of its members.[48]

Because there are no large game in the highlands, the Papua are are primitive horticulturists. Their primary crops include sweet potatoes, yams, taro, bananas, and sugarcane. They also raise pigs and chickens for meat. Although the rivers and lakes have no fish, the women net crayfish, dragonfly larvae, tadpoles, and water bugs.[49]

Several villages may be united into a tribe, either along a patrilineage or by an alliance for political reasons. A tribe consists of between several hundred and more than a thousand individuals, and the largest consist of as many as nine villages. Each village retains its autonomy that allows it to remain neutral or to abstain from a war waged by the rest of the . tribe. A tribe is led by a headman, who is generally the wealthiest man of the tribe and has a reputation for being generous in making loans to the other members of the tribe.[50] In addition to keeping peace within his tribe, the headman also represents it in negotiations with other tribes. Rather than being a chief, the headman is an individual who is more respected than the others—a *primus inter pares.*

Conflicts between tribes are generally resolved through negotiation, but if negotiation fails, a war is likely. Indeed, war is endemic among the Papuans. The cause of war between tribes is usually a dispute between individual members of two tribes, which frequently arise over a divorce initiated by a woman who has been deserted by her husband's accusations of sorcery, a murder or a theft.[50]

If the headman is unable to negotiate a settlement acceptable to the injured party, then the injured party may appeal to the headman to initiate military action, or he may precipitate a war himself by starting a fight and then persuading his friends and relatives to join him, but he is responsible for the outcome of the conflict. If he starts the war without the approval of the headman, or if his reason for the fight is deemed to be unethical, he risks being tried in a secret council of "rich men" and being sentenced to death. A death sentence is usually carried out by parallel relatives—brothers and matrilineal cousins—who ambush and kill him with arrows. If his cause is considered to be just, and he is able to secure the support of the political leaders of his tribe, war is declared against the enemy tribe.

In approaching an enemy village, the attacking party usually announces the reason for the attack and the conditions under which they will suspend the attack, but in most cases, both sides claim that they are right, and hostilities commence, usually in a "public fight." Some of the combatants wear shields that cover their backs so they can protect themselves by turning around. A public fight becomes a battle only after a man is killed.[51]

Because there is usually a limited supply of arrows, the warriors' wives collect stray arrows for their husbands during the battle, but they have an agreed upon immunity from being shot at. If a woman is accidentally injured, the guilty warrior is subject to derision and loss of prestige. He is likely to be taunted by his own people, "You better stay home and not fight; all you can do is shoot a woman." On the occasion of seeing his first public fight, Leopold Pospisil remarked,

> I could hardly believe my eyes. While the men were engaged in deadly combat, and dead and wounded warriors were being carried away, many of the women, obviously undisturbed, were quietly collecting arrows on, behind and between the battle lines, as if they were harvesting potatoes or cucumbers. Several bolder members of the "weaker" sex even climbed a hill behind the enemy's lines and, from there, shouted advice concerning the enemy's movements to their fighting husbands.[52]

Although married women enjoy protection, unmarried girls do not and are fair game for those who come to help the warriors.

At the end of a battle, the victors plunder the losers' gardens, burn their houses, and kill all the pigs they can find. After this devastation, the victors withdraw with their plunder. A warrior who has killed an enemy performs a victory dance, shouting joyful screams, and then he gives a feast called "to eat a man," in which the warrior kills and cooks a pig for his friends and relatives.

If both sides become tired of the war before there is a clear victory, and there are an equal number of dead on both sides, the parties may agree to negotiate a peace, which is formally concluded with a pig feast and peace dance. The headmen from both sides mount the roof of the dance hall to formally declare the end of hostilities, and they urge their followers to live in peace. If the fatalities on both sides are not equal, a peace may still be concluded by the side with the fewer fatalities paying an indemnity to the other side, at a rate of 180 old Kapauku cowries (cowry shells) and two fully grown pigs for the difference in the number of warriors killed.[54]

The Papua claim that they do not like war. One man explained: "War is bad and nobody likes it. Sweet potatoes disappear, pigs disappear, gardens deteriorate, and many relatives and friends get killed. But one cannot help it. A man starts a fight and no matter how much one despises him, one has to go and help because he is one's relative or one feels sorry for him.". A war may last only a few days, with only a few casualties, or it may be a prolonged affair lasting many months, with several hundred casualties and disrupting normal life.[55]

Greek Hoplite Warfare

One of the earliest forms of warfare in the western world for which we have a detailed record is hoplite warfare that was developed in the early part of the Greek classical period at the end of the eighth century BCE. During this period, the Greek citystates were almost constantly at war with one another, and warfare was treated as a kind of team sport. Nearly all famous Greeks fought as hoplites including Socrates, Plato, the poet Archilochus, and the playwright Aeschylus.

Since the citystates did not maintain professional armies, they relied on their citizens to fight, and until the fifth century BCE each soldier-citizen supplied his own equipment. Thus, only those who could afford such equipment fought as hoplites, who were usually members of the middle class. Each man was armed with a shield (*hoplon*) about a meter in diameter, made of wood with an outer bronze rim weighing between eight and fifteen kilograms, which was usually decorated with family symbols; carried a spear (*doru*) around 2.7 meters long; and wore a helmet usually made of bronze. Some carried swords as an

Figure 3.4 A Greek Hoplite Warrior[53]

auxiliary weapon as shown in Figure 3.4. The average farmer-peasant hoplite wore no armor and carried only a shield and spear, and perhaps he wore a helmet or carried a sword. A more well-to-do hoplite wore a *linothorax*—a breastplate of stitched laminated linen, sometimes covered with leather and/or bronze scales, that weighed between 22 and 27 kilograms. Some hoplites also wore greaves, leather shin armor.

The spearhead was usually carved in the shape of a pointed leaf, and the rear end had a spike called a *sauroter* (lizard-killer), which was used to stand the spear in the ground. The *sauroter* was also used as a secondary weapon if the spear shaft broke, or for the rear ranks to finish off fallen opponents as the phalanx advanced over them.[56]

Because these citizen-soldiers, especially farmers, needed to return to their regular occupations, campaigns were of limited duration, usually lasting only a day, but they were bloody and brutal. The battles were usually arranged beforehand and were usually fought on a flat, open plain. The warriors were arranged in phalanxes, usually eight men deep. The hoplite carried his shield on the left arm that protected not only himself but also the man to his left. This left the right flank of the phalanx only half protected, and each side tried to exploit this vulnerability by attempting to overlap its enemy's right flank. Because of this, the phalanx tended to drift to the right during the battle because the hoplite would try to stay behind the shield of the neighbor on his right.

Hoplite warfare was shock or short-range combat. The two phalanxes would smash into each other at a walking pace, hoping to break or encircle their opponents. Failing that, the battle degenerated into a pushing match in which the rear would try to push the front line into the enemy and cause it to break ranks and disperse. Thus, the side with the deeper phalanx or with more men had an advantage. Battles rarely lasted more than an hour because of fatigue, and once one of the lines broke, the troops would flee from the field, chased by *paltasts* (light infantry) and they would sometimes drop their heavy shields, disgracing their friends and families. Causalities were usually light, rarely exceeding 5 percent. Victory was enforced by ransoming the fallen back to the losing side, a process called the Custom of the Greeks.[57]

Greek hoplite warfare is in some ways a forerunner of modern warfare because of its dependence on shock, the requirement of disciplined soldiers, the high degree of training required and the organizational structure of the hoplite army.

Nonviolent Societies

The Xinguano Indians. Although most primitive societies engage in some form of warfare, there are a few instances of peaceful, non-warring societies. However if such societies are attacked, they usually defend themselves. One example of such a peaceful society is the Xinguano peoples in Brazil's upper Xingu River basin. There are approximately 1200 Xinguano Indians living in ten tribal villages that are divided into four language groups. Each village remains a separate and politically autonomous community with a strong sense of its uniqueness. A central feature of the Xinguano culture is the concept of peacefulness, and their concept of goodness is to remain peaceful. A good man avoids confrontations and does not show anger. He refrains from injuring others because he would sense their pain as if it were his own. The Xinguano believe that a violent person lacks this sensitivity and kills and maims others, often without a motive. The reason that peace makes for good policy is because, "the peaceful man is treated peacefully by others; after death, only the peaceful man lives in the paradisiacal Village in the Sky."[58]

There is considerable intermarriage among the Xinguano tribes. If a man cannot find a suitable mate in his own village, he may marry into another village and then move into the bride's village. Approximately one-third of the marriages are of this kind. The

Xinguano are one of the few human primitive societies in which males leave their natal group. The movement of males between villages tends to break down the bonds that bind related males together in patrilocal societies, which in most societies contributes to militarism and a willingness of males to fight for their clan. Xinguano villages are not matriarchies, their chiefs are males who are regarded more like fathers to the other members of the community than as political leaders in the usual sense. The chief's primary political role is to mediate disputes between males of different villages who chop down trees on the paths between villages, and who "borrow" another villager's canoe or bicycle.[59] The Xinguano have been the victims of unprovoked aggression and raids. One raid, which occurred in the 1960s, in which Xinguanos were shot and kidnapped and their communities ransacked, precipitated a violent defensive reaction by the Xinguano in which many of the raiding tribe's men, women and children were killed. The victorious Xinguano men took no trophies and received no special honors. After the war, they were forced to take special potions to rid themselves of the enemy's defiling blood before they could reenter their villages. Such warfare is regarded as an exception to the peacefulness of the Xinguano.[60]

The Semai. Another example of a normally peaceful group of primitive humans that has been stimulated to fight is the Semai. The Semai were swidden agriculturists on the Malay Peninsula who have a long reputation of being a peaceful people. They have a taboo against anger and have a collective self-image of themselves as being nonviolent. Instead of saying, "Anger is bad," they say, "We do not get angry," and an angry man will deny his anger.[61] However during the communist insurgency in Malaya during the early 1950s, the Semai were reduced to dire economic straits, and the British recruited troops among them to fight the communist insurgents.

The Semai were shocked to learn that as soldiers, they were expected to kill other men, and many people who knew the Semai insisted that they would never make soldiers, but after the communists killed some of their kinsmen, they wanted revenge and became enthusiastic warriors. Anthropologist Robert Dentan reported: "Taken out of their nonviolent society and ordered to kill, they seem to have been swept up in a sort of insanity they called 'blood drunkenness.'" A typical veteran's story runs like this: "We killed, killed, killed. The Malays would stop and go through people's pockets and take their watches and Money.

Table 3.1
The Causes of Primitive Warfare

Motive	Western Indians	World
Revenge, retaliation & defens*	75%	94%†
Economic (bounty, territory, pouching,)slaves)	85	70
Women	n.c.	58
Personal (prestige, trophies, visions)	44	36
Political (subjugation, tribute)	n.c.	6

* includes defense of territory
† Percentages are for all societies reporting these as causes or a motive for war
n.c. - Not counted

We did not think of watches or money. We thought only of killing. Wah, truly we were drunk with blood."

After the war, the Semai soldiers seemed bemused, but not displeased, that they had been such good soldiers, but they were unable to account for their behavior. It was as if they had put the experience in separate compartment in their minds. When they returned to their traditional society, they were as gentle and as afraid of violence as any of their kinsmen.[62]

The Nature and Causes of Prehistoric War

Although a few primitive tribes such as the Papua New Guinea and the Australian Aborigines fought in battles that involved more than a few individuals, the preferred form of warfare for both primitive humans and chimpanzees was the raid in which a few warriors would attack and kill a single individual. A raid was considered to be a failure if a single warrior was killed because the death of a single warrior would be a major loss for a small hunter-gatherer band. It was not until human societies grew to more than several thousand or more individuals that it could afford to sustain the kind of losses that are involved in modern conventional wars.

The fact that the raid was the preferred form of warfare for both chimpanzees and primitive humans supports our hypothesis that warfare originated with the common ancestor of humans and chimpanzees and was inherited by both species from their common ancestor.

The primary causes of primitive warfare are revenge and retaliation for previous injuries, insults, or theft;[63] to provide an opportunity for males to demonstrate their courage and manliness; and for some territorial societies, to defend territory. In many societies, an aggrieved individual organizes a raid and recruits its participants. Primitive humans seldom engage in lethal conflict for political reasons to conquer or destroy their enemies. For them, warfare appears to be an end in itself and is considered to be an essential part of the life of a male. Warriors become heroes, and those killed in battle are highly honored

Table 3.1 shows the results of two ethnographic surveys of the motives and causes of primitive warfare. The first, by Keith Otterbein, includes 50 pre-state societies,[64] and the other, by Joseph G. Jorgensen and compiled by Lawrence H. Keeley, includes 173 western North American Indian tribes.[65,66] Because defense in the first category of the chart includes defense of territory, a very high percentage of these wars was likely fought for territorial reasons.

Warfare and Natural Selection

Because warfare can result in large numbers of human fatalities of both warriors and noncombatants, it might be thought that warfare decreases fitness of both warriors and noncombatants and that natural selection would have removed any genes responsible for warfare from the human gene pool. However warfare among primitive societies and chimpanzee groups are usually raids in which fatalities are very low, and the raid would be considered a failure if a single member of the raiding party was killed. Among the Yąnomamö Indians, approximately 30 percent of all male deaths are the result of homicide, approximately 40 percent of adult males have participated in killing another person, and some have participated in more than one killing. However men who have killed—*uhnokais*—are sexually more successful than men who have not killed. On the average, *uhnokais* have two and a half times as many wives and three times as many children as other men.[67] Among the Plains Indians, a man could not marry until he had earned the right to, which, for the Crow Indians, meant that a man either had *counted coup* or had reached the age of 25.[68] Thus, from an evolutionary viewpoint, warfare seems to have been the product of natural selection because of the greater fitness of warriors of prehistoric societies. Charles Darwin recognized this when he wrote in *The Descent of Man*,

> "There can be no doubt that a tribe including many members who, from possessing in a high degree the spirit of patriotism, fidelity, obedience, courage, and sympathy, were always ready to give aid to each other and to sacrifice themselves for the common good, could be victorious over other groups *and this would be natural selection.*"[69] (italics added)

While war appears to have enhanced the biological fitness of individual warriors, after the agricultural revolution, it also enhanced the prosperity and average fitness of the victorious community as a whole.

It was not until humans began to live in large societies and to fight modern wars with the objective of conquering or even eliminating the enemy, between 7,500 and 5,000 years ago, that fatality rates increased to the point that wars began to become counterproductive from an evolutionary point of view. Yet fatality rates remained below the levels that would lead to extinction of the human race. However some wars such as the Peloponnesian Wars (431 – 404 BCE) and the Thirty Years War (1618 – 1648 CE) in which the population of Athens and Bohemia were either eliminated or dispersed, changing the population distributions and thus causing natural selection to displace the weaker populations with the more warelike populations.

The Myth of the Peaceful, Noble Savage

There is a tradition in western philosophy and in anthropology of the peaceful, noble savage, which posits that our primitive ancestors lived in peaceful cooperative societies in which warfare and other forms of violence and aggression were absent, and that man's violence emerged only when humans became civilized. The French Enlightenment philosophe Jean-Jacques Rousseau (1712–1778) wrote that human nature was basically peaceful but became aggressive and warlike when man became civilized. Rousseau believed that war is a product of civilization, that it is an institution of private property that has alienated humans from their original freedom and happiness and has corrupted them by creating inequality, a class system, and laws based on force:

"Nascent society gave way to the most horrible state of war; the human race, degraded and devastated, unable to retrace its steps or give up the unfortunate acquisitions that it had made. . . brought itself to the brink of ruin."[70] In the end, however, Rousseau admitted that his ideal "state [of nature] . . . no longer exists, or may never have existed, and will probably never exist in the future." Rousseau's nascent society was not a fact, only a standard of compression: "Let us begin, then, by setting aside all the facts, for they are irrelevant to the problem. The investigation that may be made concerning this subject should not be taken as historical truths, but only as hypothetical and conditional reasoning's, better suited to casting light on the nature of things than to showing their real origin."[71]

Contemporary advocates of this point of view include David Fabbro, Raymond Kelley, and Keith Otterbein. Fabbro was able to find seven societies that meet his eight criteria for a warless society.[72] Although his list includes both the !Kung and Copper Eskimos, which are currently considered to be peaceful by some ethnologists, both of these societies had engaged in small scale raids and prolonged feuds before they were pacified by the Bechuanaland/Botswana Police and the Royal Canadian Mounted Police, respectively.[73] These two societies thus must be classified as warlike according to our definition of war: "socially sanctioned, organized lethal intergroup conflict." Fabbro also includes the Semai of Malaya, who had fought as soldiers for the British during the communist insurgency in the 1950s. Furthermore, the members of all of Fabbro's warless societies engage in interpersonal violence in one form or another.[74]

Raymond Kelly found that out of eight primitive foraging bands, six rarely or never engaged in warlike activities;[79] however he also included the !Kung and the Copper Eskimos in his list of peaceful societies. He concludes, "Warfare is typically rare to nonexistent within and between unsegmented foraging societies inhabiting environments characterized by low population density, diversity, and predictability at densities below 0.2 persons per square mile."[76] He concludes his study by observing that war and society (civilization?) coevolved and that the transition from primitive warfare to modern warfare occurred after 10,000 years ago, at the beginning of the agricultural revolution.[77]

Keith Otterbine believes that prehistoric humans may have fought wars but only rarely and that warfare developed among hunter-gatherers when they developed weapons and began hunting big game, and that the incidence of warfare declined during the early part of the agricultural revolution coincidently with the decline of big game hunting and the domestication of animals. Otterbine also claims that warfare was not prevalent during the formation of states during the latter part of the agricultural; revolution. Modern warfare began after the emergence of states. and the institution of armies.[78] (But why did armies emerge?) As we shall see in the next chapter, the preponderance of evidence indicates warfare was a major factor the in formation of states. In the chapters that follow we will explore some of the factors that led humans to be so warlike.

Chapter 4

The Origins of Civilization and Modern War

> From the very beginning the conquest of one people over another has been, in the main the conquest of social man over antisocial man; or, strictly speaking, of the more adapted over the less adapted.
>
> Herbert Spencer[1]

The Agricultural Revolution

The agriculture revolution that originated in Southwest Asia between 13,000 and 10,000 years ago was one of the most important events in human evolution because it provided humans with the conditions necessary for the development of civilization and was the demarcation between prehistory and civilization. It also marked a dramatic change in the nature of war because war played an essential role in the creation of states and civilizations.

During the Pleistocene epoch (the last ice age), much of the earth's water was trapped in a vast ice sheet that covered much of the northern hemisphere north of about 50 degrees latitude, including England, Scandinavia, Northern Germany, Poland, and most of Russia in the eastern hemisphere and Canada in the western hemisphere. Ocean levels were about 120 meters below their present levels. South of the ice sheet, the northern hemisphere was a cold, arid grassland similar to the tundra of the subarctic regions today. This climate supported large populations of herbivores (plant eating animals) including reindeer, horses, woolly mammoth, bison, woolly rhinoceros, cave bears, and wolves that provided a rich protein diet for the Paleolithic (old Stone Age) man, who favored a meat diet because it was rich in fats, which provide warmth in this cold climate.

When the earth began to warm about 12,000 years ago at the end of the Pleistocene epoch, the ice sheet began to melt and large quantities of water were released into the atmosphere that fell as rain and snow, giving way to the current Holocene epoch with its warmer, wetter climate. As the ice sheet retreated north, the large animals that had provided food for humans migrated northward into areas that became accessible to them making it harder for the humans to feed themselves, so humans began to rely more on fruits, nuts, and tubers that they were able to gather. Recent archeological discoveries indicate, however, that seasonal variations in the rainfall during this period may have reduced the nutritional intake of the foragers by as much as 50 percent during draughts, so that many humans may have suffered seasonal or even long-term starvation, thus keeping human population densities low.[2]

Every environment has a limit to the size of the population it can support, its carrying capacity. Tropical woodlands can support far more individuals than deserts and arctic regions can. Furthermore, a carrying capacity depends on the available technology. Land farmed with modern agricultural technology can support many times the number of people that it can support by hunting and foraging. If the population exceeds the carrying capacity of its environment, either the population must decrease to the carrying capacity, or a new technology must be introduced to increase the land's carrying capacity. Population size can be controlled in several ways. A community can adopt birth control, or it can practice infanticide as some primitive societies have done. For hunter-gatherers, this has meant low population densities, particularly in harsh climates such as deserts and the arctic.

About 13,000 years ago in northern Syria and 12,000 years ago in northern Iraq, human foragers found stands of wild forms of emmer and einkorn wheat. Beginning about 11,500 years ago, humans began cultivating wheat, barley, chickpea, lentil, flax, and pea in the Fertile Crescent that extends from northern Syria across northern Iraq and south to the foothills of the Zagros Mountains in eastern Iran.[3] Because these crops provided humans with an adequate diet of carbohydrates and they could hunt small game such as goats, pigs, and gazelles that remained in the region, humans began to settle permanently in small villages that grew in size up to several hundred people/ They built houses of mud brick, stone and wood, and began to cultivate crops.

These early farmers discovered that if the grains and cereals were cultivated and irrigated, and if other plants—weeds—were removed to keep them from choking their food plants and from using up precious water and soil nutrients, the yields were much higher than those of the wild plants.[5] By using the seeds from the plants that produced the highest yields and the largest grains as crop seed, humans discovered how to improve their crop yields and thereby induced the evolution of their crops through artificial selection. Over the next 3,000 years, the genetic composition of these wild grains evolved and changed so they could no longer grow wild but required cultivation.[6, 7]

When these early farmers began domesticating animals about 9,500 years ago,[8] they discovered that their animal stock could be improved by selecting the largest and most productive animals for breeding. Sheep were breed from a goat-sheep ancestor to produce the familiar thick woolly coats that could be shorn, spun, and woven into cloth.

Because plants grow seasonably, these early farmers had to devise ways to store and preserve the food they produced during the growing seasons, and they learned that grains and cereals were the best bets for long-term storage. At first they stored their grains in underground pits, but food stored in pits deteriorates rapidly and cannot be stored for more than a winter, and would not last through a lean year, so they innovated baskets and pottery for longer-term storage.

This transition from foraging to farming did not happen overnight. In some locations people practiced both farming and foraging before giving up foraging. Some contemporary nomads in New Guinea plant bananas and papaya, go off on long hunting trips, and then return to harvest their crops. The Apache Indians in the southwestern United States settled in villages in the highlands, farmed in the summer, and moved to the southern lowlands to hunt game in the winter.[9] In other places such as coastal Peru, in the river valleys of the eastern United States and the Columbia River delta, hunter-gatherers established permanent settlements long before they took up agriculture. These locations provided an ample year-round supply of wild fruits, nuts, tubers, and fish so that seasonal migration was not necessary.

The agricultural revolution also resulted in changes in man's lifestyle and was the genesis of civilization. During the four or five millennia that followed man's first agricultural endeavors in the Fertile Crescent, agricultural was transformed from a subsistence endeavor into an industry. Agricultural technology slowly improved as domesticated animals were harnessed to plow the land and as irrigation, milling, and food storage systems were developed. Because farming resulted in a much greater carrying capacity of the land than foraging and hunting did, the populations of these agricultural settlements began to grow at an exponential rate. Mothers substituted cow's and goat's milk for mother's milk, so the interval between births decreased from four or five years to between one and two years, and consequently, birth rates doubled. Adults began to eat cheeses and yogurt and developed a tolerance for lactose that they had previously lost in early childhood when they were weaned from mother's milk.

But farming had its downside: diets usually included a smaller variety of foods, which did not provide as healthy a diet as that of hunter-gatherer ancestors. Furthermore, living in large permanent settlements enabled diseases to develop and spread because of lack of clean drinking water, the absence of latrines that resulted in the contamination of drinking water, and reduced breast-feeding as maternal fecundity increased.[10] People also caught the diseases of their domesticated animals, such as smallpox and chickenpox, and pathogens from the animal milks that were substituted for mother's milk. Tooth decay and abscesses were common. Although birth rates may have doubled, populations grew more slowly because of these new diseases.

The Emergence of Civilization

The increased agricultural efficiency resulted in food surpluses that allowed some individuals to divert their endeavors from agriculture to other occupations, such as craftsmen, builders, traders, priests and soldiers; and this diversification of labor was an important result of the agricultural revolution and was the beginning of civilization. During this period, the size of agricultural communities grew from small settlements into villages of perhaps fewer than 500 residents that were egalitarian with decisions made collectively, then into towns of up to 50,000 residents that required greater organization and more formal governance than did small villages.

As these agricultural communities grew and personal relationships between their members became more complex, they had to learn to live with non-kin strangers, the need for governments emerged. Not only did law and order have to be established and maintained, but labor had to be mobilized to build irrigation works, roads, fortifications, palaces, and temples. This new form of governance was frequently led by chiefs who were dominant men able to organize the activities of the community, make and enforce laws, and organize defenses for the community. In order to maintain order and to organize community efforts, the ability of the chiefs to maintain a monopoly on the use of force developed and became a permanent institution. The chiefs made all the significant decisions for the community, and in some communities, rudimentary bureaucracies emerged.[11] At the beginning of the agricultural revolution, when the communities remained small, women continued to migrate from their natal groups into neighboring groups to avoid genetic inbreeding. But as the agricultural communities grew, it was no longer necessary for females to migrate to avoid inbreeding, so females remained in their villages and towns and married local men. Thus there were fewer opportunities for social contacts and friendships to develop between the communities, which made it easier for hostilities to develop between the communities.

These early farmers were constantly at the mercy of the weather. Some years produced bountiful crops, whereas other years produced meager ones and even crop failures, either because of the lack of rain or because of pests such as locusts, so that these early societies were in a constant life and death struggle with their environment. To explain these life threatening events, they invented supernatural gods and spirits, which they believed would relieve their sufferings if they placated them with offerings and sacrifices, and in some places, the gods demanded human or human blood sacrifices. Shrines and temples were built to house the images of the gods. In some places, the pantheon included thousands of gods, and as the number of gods multiplied, a priesthood emerged to care for the idols and to mediate between the people and their gods. Because of the mystical powers claimed by the priests, in some places they became the rulers of their communities. The temples became the administrative centers for the communities, the storage places for the cities' wealth, and the dwelling places of the priest-kings, who levied taxes with which they paid themselves and the temple staffs and built more temples and shrines to the gods and spirits. They employed shepherds to tend the temple flocks and craftsmen to build the temples and make the sacred objects that filled them. The temple sites also served as markets and bazaars for the barter and exchange of goods.

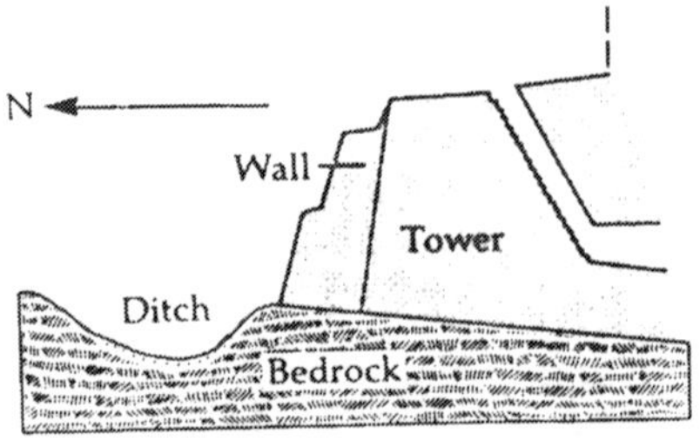

Figure 4.1 – Cross-section of the fortification at Jericho.[12]

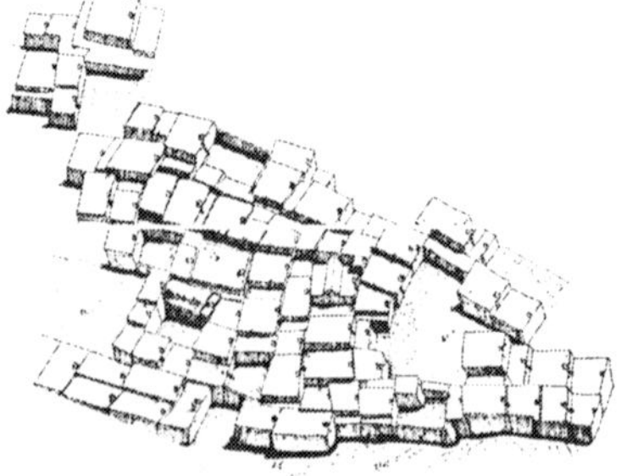

Figure 4.2– Neolithic village at Catal Hűyűk in southeast Asia.[13]

Religion became a major factor in promoting and facilitating the profound social and psychological changes that were required by the agricultural revolution and urban living. The populations shifted their allegiances from their kin groups to the larger groups, and the priesthood developed an organized system of beliefs that became integrated into the social order through public rituals that linked religion to the political and military objectives of the state. Thus religion became an instrument of the state. In Egypt the pharaohs were deified and their military achievements were perceived as divinely inspired or ordained.

The Emergence of Modern War

Not all human communities adopted agriculture simultaneously. Those that did prospered and became the envy of those that retained their hunter-gatherer way of life and they began raiding the agricultural communities. Thus the need arose for the agriculturists to defend their communities and they began to organize defenses for their settlements and and their farms. By that time, long-range weapons such as the bow and arrow, the sling, and the atlatl (spear thrower) were in common use, and as a result, some of the settlements developed fortifications. One of the earliest of these was the Neolithic settlement at Jericho, an oasis site in the Jordan River valley that existed between 8350 and 7350 BCE and was surrounded by a wall three meters thick and four or more meters high, as shown in Figure 4.1. There was a tower ten meters in diameter and nine meters high at one corner. Because the Jerichoians were archers, the tower and walls gave them an advantage over attackers The walls enclosed an area of about ten acres.[14] Another fortified city, Catal Hűyűk, existed in southern Anatolia between

9,000 and 8,300 years ago. Its population of between 5,000 and 6,000 lived in dwellings as shown in Figure 4.2, which were built so that the outer walls of the dwelling formed a continuous wall that presented a defensive barrier to attackers. If an attacker managed to break through the outer wall, he would find himself not inside the town but in a single room. Since the entrance to these dwellings was by a ladder through an opening in the roof, the ladder could be with-drawn, and the attacker would be trapped and could be engaged from the roof.

Figure 4.3 – Victory Stele of Nuram, king of Akkad (The Louvre)

In Mesopotamia, Neolithic raiders from Akkad in the north and Elam (ancient Persia), seeking tempting loot from the towns, forced their residents to fortify their towns for protection. Several wall murals (e.g. Figure 4.3) shows warlike scenes such as the wholesale killing of unarmed captives after a military victory.[15] The conflicts between the Sumerians and the Elamites likely extended back to Neolithic times, but the first records of war between them appeared in 2600 BCE, when Men-barage-si, the first king on the Sumerian King List, undertook a war against the Elamites and "carried away as spoils, the weapons of Elam."[16]

War and the Emergence of the State

The historical and archeological records show that war played a major role in the formation of the early states in Mesopotamia, Egypt, India, China, Japan, Rome, northern Europe, Mesoamerica and South America.

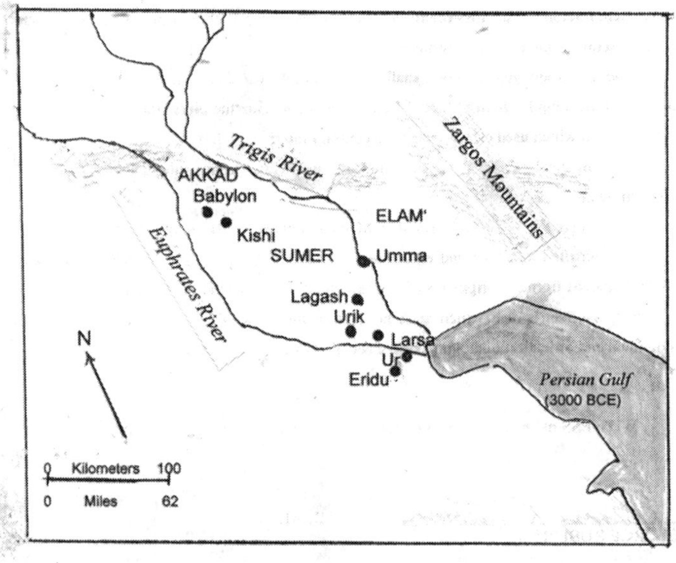

Figure 4.4. Ancient Mesopotamia.

Anthropologist Robert L. Carneiro has argued that the first agricultural societies in Mesopotamia, the Nile and Indus valleys, China, Mexico and the coastal valleys of Peru were formed by war and conquest in places where the agricultural land was geographically circumscribed by mountains, seas or deserts which sharply limited the land that could be farmed.[17]

Uasing Mesopotamia as an example, the first agricultural settlements in southern Mesopotamia date from about 7,300 years ago at Eridu. A millennium later, as many as 28 independent citystates and agricultural settlements arose on the alluvial plane of southern Mesopotamia as shown in Figure 4.4 some of which had populations as large as 10,000 inhabitants. At the center of each town or city was a temple dedicated to the city's patron god or goddess. These early Sumerian cities showed a high degree of social and economic diversity in the form of artisans, merchants, priests, and bureaucrats. Most were theocracies ruled by priest kings assisted by councils of elders that included both women and men.

As the populations of these towns grew, irrigation ditches were dug to irrigate more land, but because the soil absorbed so much of the water in the ditches, they could be extended only a few kilometers from the rivers, thus limiting the arable land. It was during the Uruk period (4100–2900 BCE) that the Sumerians began to use slaves captured in the hills of

Elam, but there is little evidence of organized warfare between the citystates during this period. The Uruk period ended with the end of the Piora Oscillation (3200–2900 BCE) that ended the period of a wet, cooler climate that had facilitated agriculture in Mesopotamia.

By the beginning of the fourth millennium BCE, some of these towns had grown into cities that held as many as 50,000 people, and all of the arable land had been brought under cultivation. By 3300 BCE, Uruk covered an area of three square kilometers (175 acres) and had a population of as many as 40,000 inhabitants. As the citystates grew and became circumscribed by the surrounding desert, disputes among them arose over land and water rights, and they began fighting over these rights.[18] By the middle of the third millennium BCE, there was almost constant warfare among the citystates, so that warfare increasingly dominated Sumerian life.

These wars were fought for economic reasons, and their intensity increased. The first armies are believed to have emerged in Mesopotamia. Initially, they were "city guardians," but the citystates became engaged in almost continuous warfare. Several wall murals show warlike scenes such as the wholesale killing of unarmed captives after a military victory as shown in Figure 4.3.[19] The larger towns challenged the smaller ones, and the defeated towns either were eliminated, expelled from the region, or became politically subordinated to the victor and forced to pay a tribute to their conqueror.

According to the epic literature of Sumer, Gilgamesh, the legendary king of Uruk, circa 2600 BCE, challenged the power of the chief priest and became the king of Uruk sometime during the first half of the third millennium BCE. He built a massive defensive wall around the city and defended it against a siege by his rival Agga, the ruler of Kishi, who he defeated.[20] There is evidence of serious warfare throughout Mesopotamia over the next two centuries until Lugal-kinishe-dudu (r. 2371–2296 BCE) was able to unite all of Sumer and became king of Kishi. He concluded a nonaggression pact with Entemena, king of Lagash. As towns conquered villages and cities conquered towns, the size of these polities grew until the entire region was united into the states of Sumer, Elam, and Akkad as shown in Figure 4.4. In 2316 BCE, the entire region was united into a single state under Lugal-zage-si (2340–2316 BCE), the ambitious ruler of Umma. He is said to have looted, burned, and destroyed practically all of the holy places of Lagash, and he claimed to have conquered all of the land from the rising sun to the setting sun, and all of the people from the lower sea (the Persian Gulf), along the Tigris and Euphrates Rivers to the upper Sea (the Mediterranean Sea), "the people of all the lands lie peacefully in the meadow under him; the land rejoiced under his rule; all the chieftains of Sumer and the rulers of all the foreign lands bowed down to him in Erech in accordance with the rules of princeship."[21] Thus, Sumer became the first large integrated state and also the first civilization.

In addition to the geographic circumscription described above, Napoleon Chagnon has introduced the concept of social circumscription, which is the tendency for villages to clump together even though there is no geographic circumscription. Chagnon noted that in the upper reaches of the Amazon River, approximately 10,000 Yąnomamö Indians live in closely spaced villages to the extent of impinging on one another, in spite of the fact that there is ample land available in the upper Amazon for the villages to be spaced further apart. The result is that the villages in the center engage in frequent and intense warfare, whereas those on the periphery, where the villages are spaced further apart, fight much less frequently. Moreover it is more difficult for villages in the center to avoid attack by moving away than it is for villages on the periphery. The villages in the center tend to be larger than those on the periphery because larger villages have an advantage in war. Furthermore in addition to geographical and social circumscription, competition for resources also results in the intensification of warfare.[22]

Military Technology in Sumer

The almost constant warfare among Sumer's citystates during the third millennium BCE stimulated the development of military technology and techniques far beyond those of the primitive societies that existed in prior times. The first war for which there is any detailed record is the war between the citystates of Lagash and Umma in 2525 BCE, in which Eannatum, the king of Lagash, defeated Umma. Afterward Eannatum erected the Stele of Vultures (Figure 4.5) to commemorate his victory. The stele portrays Eannatum leading a phalanx of armored and helmeted infantry warriors armed with spears, trampling on their enemies. Eannatum, armed with a socketed axe, rides in a four wheeled chariot drawn by four onagers (wild asses). In the lower panel of the stele, Eannatum holds a sickle sword. The stele is the first known historical document. On the reverse side is a description of the battle. It provides the first evidence of a standing professional army. It also provides the first

Figure 4.5 – The Stele of Vultures (The Louvre)

evidence of the use of copper helmets, that probably had leather liners. They were found in the Death Pits of Ur dating from 2500 BCE. The introduction of the helmet was probably a defensive response to the killing power of the mace, probably one of the oldest weapons, and was extremely effective against a soldier who had no head protection. Thus, the helmet was a major advance in military technology. The stele also shows the soldiers wearing armored cloaks, probably made of thin leather and covered with metal disks with raised centers. Although somewhat primitive, these cloaks were the first appearance of body armor, which provided some protection against the weapons of that time. Later, the Sumerians introduced overlapping metal plate body armor.. The Sumerian invention of the chariot ranks among the major military innovations, but the chariot probably was not an effective offensive weapon because of its size, weight, and its instability. For the late Bronze and early Iron Ages, however, the chariot became the primary striking weapon in the hands of Hittites, Canaanites, Egyptians, and Assyrians. Charioteers, archers, and spearmen were the elite fighters in the ancient world, and in some places, this tradition continues to this day. The Israelis named their first tank the Merkava—the chariot. The Sumerians also used the wheeled cart for logistical transport, which became the standard logistics vehicle in the Middle East until the time of Alexander the Great.

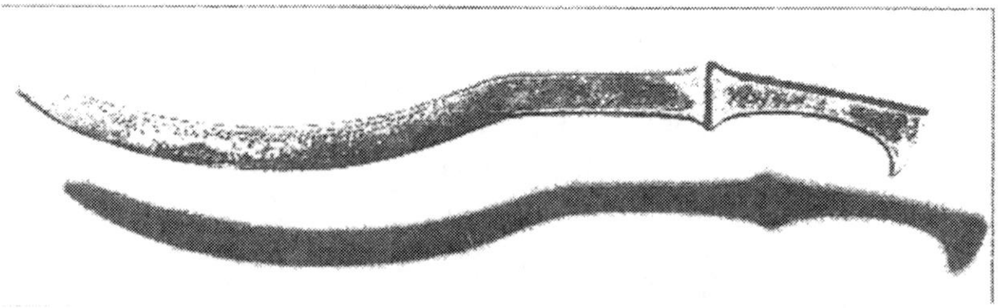

Figure 4.6 - The Sickle Sword (Assyrian, 13th Century BCE).[23]

The two major weapons introduced at the time were the socketed axe and the sickle sword. The cast bronze blade of the socketed axe was slipped over its shaft and held in place with a pin. In response to the development of body armor described above, significant modifications to the socketed axe were introduced about 2500 BCE. The blade was narrowed to concentrate the impact, and the socket was strengthened to penetrate the bronze plate armor that had been introduced at that time. The socketed axe remained in use for 2,000 years.

The sickle sword shown in Figure 4.6, was so named because it is shaped like a sickle. It was introduced by the Sumerians about 1300 BCE and became the primary infantry weapon of the Egyptians and biblical armies at later times. When the bible speaks of people being "smote," it refers to being attacked with the sickle sword.

The composite bow, made of wood, horn, and sinew, was introduced during the reign of Sargon's grandson Naram-Suen (r. 2291–2255 BCE). The composite bow was stronger and more resilient than the simple wood bow. Whereas the simple bow had an effective range of between 50 and 100 meters, it could not penetrate simple armor at these ranges. The composite bow had a pull two to three times that of the simple bow and a range twice that of the simple bow, and it could penetrate leather armor at that range. It was in use by all armies for the next fifteen centuries.

All the major citystates of Sumer and Akkad maintained standing armies of between 600 and 700 soldiers who served as the king's bodyguard and formed the professional core of the army. These forces could be augmented to between six and seven thousand fighters by conscription from the male populations. The smaller cities would have contributed their armies to those of the larger citystates when they went to war. Another source of military manpower were captives from non-

Sumerian cities. The armies of ancient Egypt, Assyria, Persia, the Greek citystates and Rome all contained large contingents of their former enemies.

Universal State Formation

The process of state formation and the emergence of civilizations described above was repeated again and again in Egypt, Syria and Anatolia, and later in China, Europe and in the western hemisphere. In almost all cases, war played a major role in the development of these civilizations.

Egypt. The development of the Egyptian civilization in the Nile Valley followed a pattern similar to that in Mesopotamia The Nile flows through a narrow valley bounded on both sides by deserts so that agriculture was confined to the valley. The river floods annually from monsoons at the river's sources in equatorial Africa and the Ethiopian highlands, and as the floods receded in the late spring a rich top soil was deposited along its banks. From present day Khartoum where the Niles two main tributaries—the White Nile and the Blue Nile—join, to Elephantine, the Nile flows over six cataracts or rocky narrows. Circa 4000 BCE, agricultural settlements developed along the river's banks to take advantage of the fertile soil. The Nile not only provided fertile soil for growing crops, but it also attacked game such as water buffalo that could be either killed for meat or captured, tamed and used for plowing The settlements eventually grew into towns, each ruled by a chief or monarch, and over time these towns united into two loose federations of Upper Egypt and Lower Egypt. Although these regions were never clearly delineated, Lower Egypt seems to have consisted of the delta .region and the area immed-

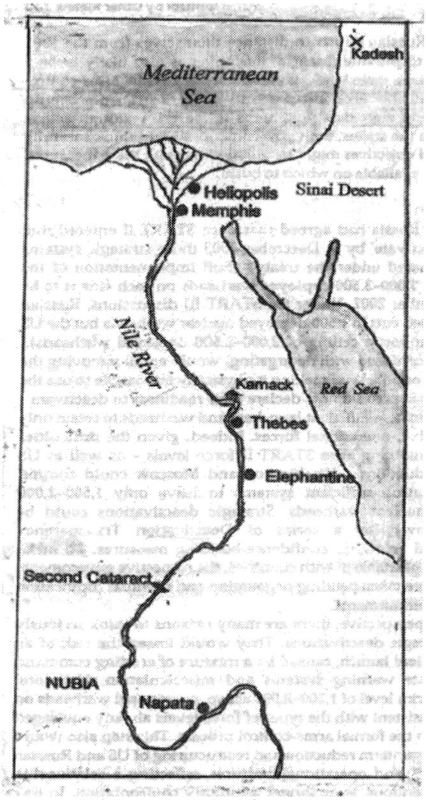

Figure 4.7 Ancient Egypt

iately south of its apex. Upper Egypt seems to have extended southward from that point to the first cataract south of Elephantine Then around 3150 BCE the ruler of Heliopolis conquered lower Egypt and built a new capital at Memphis just south of the delta, thus establishing an Egyptian state that extended from the second cataract to the Mediterranean Sea as shown in Figure 4.7. This first king of ancient Egypt has been variously known as Narmer, Memes and Hor Aha (Fighting Hawk). Power shifted northward, creating tensions between Upper and Lower Egypt which were to resurface intermittently throughout Egypt's history.[24]

The half millennium following Namur is known as the early dynastic period (c. 3100-2613 BCE). Because the deserts provided Egypt with natural defenses to the east and west of the Nile and during that period, the Nubians in the south did not pose a threat. to Egypt, it only had to contend with occasional uprisings of local populations. During this period, the Egyptians established trading relations with the Nubians to the south and with towns along the Mediterranean coast to the north.[25]

During the Old Kingdom (2613–2160 BCE), Egypt established its first regular army, which undertook campaigns in the Sinai, against the Libyans to the west, and the Nubians to the south, where it captured large numbers of slaves and cattle and exploited Nubia for its gold and copper mines and other minerals.

But the real threats to the Old Kingdom were the constant rebellions by local chiefs who retained their own armies and exercised control over the strategic trade routes along the Nile. Thus a national army was needed to deal with these periodic revolts. The pharaoh's army consisted of several thousand troops organized as a household guard that was augmented by conscription in which one man in a hundred was called to service each year. At its outer reaches, however, the central government depended upon regional governors, some of whom became independent monarchs who organized their own armies and founded their own dynasties. Around 2200 BCE Egypt was devastated by a series of agricultural failures due to the desiccation of the climate, and the central government lost its control over the country. A civil war broke between Upper and Lower Egypt, which Lower Egypt eventually won around 2050 BCE and reunited the country under the Theban ruler Mentuhotep II.

The reign of Mentuhotep II (r. 2046 – 1995 BCE) marked the beginning of the Middle Kingdom, an era of peace and stability in Egypt during which the country grew wealthy through trade and conquest. Egypt conquered Nubia around 2000 BCE and built a series of fortified towns, the largest of which was Buhen just below the Second Cataract. The

Egyptians established a trading center at Byblos on the coast of the Levant that had a thriving colony of Egyptian merchants. Trading routes into the Levant were protected by military units. Copper mines were developed in the Sinai.

About 1750 BCE, the Middle Kingdom was overthrown by the Hyksos (Canaanites from the hills of southern Jordan). The Hyksos had built fortified towns along the Egyptian trading routes in the Levant. They invaded Egypt and overthrew the central government, and brought with them bronze metallurgy, the laminated bow and the chariot which were previously unknown to the Egyptians. These new weapons were superior to those of the Egyptians and gave the Hyksos the advantage that allowed them to conquer the delta region, where they established their capital at Avaris. The Egyptians retained control of the Nile Valley south of the delta. In the middle of the 16th century BCE, Pharaoh Ahmose I drove the Hyksos out of Egypt.[26]

Ahmose I's reign marked the beginning of the New Kingdom, which was led by a series of warrior pharaohs, who developed powerful armies that extended Egypt's borders to include the Sinai and the Levant up to Anatolia and as far south as Napata, as shown in Figure 4.7. In the north, Tutmose I (r. 1504–1492 BCE) led his armies to the Euphrates River in Syria. His victory against the Mitanni at Megiddo circa 1460 BCE gave Egypt control of the entire Levant. The peace treaty included the marriage of a Mitanni princess to Pharaoh Amenhotep III. Because Egypt had few trees, it imported cedar lumber from Lebanon. Circa 1400 BCE, during the reign of Queen Hatshepsut, the Egyptians established a port at Punt on the Horn of Africa.[27]

Egypt's control of the Levant was challenged by the Hittite Empire, centered in Anatolia. The Egyptians, led by the last of the warrior pharaohs, Ramses II, claimed a victory at the chariot battle at Kadesh in 1300 BCE, but in 1265 BCE, Ramses II was forced to cede control of northern Levant to the Hittites. Ramses III, however, rallied Egypt's military strength and in 1186 BCE defeated the "sea people," a confederacy of pirates in the eastern Mediterranean Sea who ravaging Egypt's coasts. By that time, however, the New Kingdom was in a state of decay, and by 1100 BCE, both Nubia and the Levant had been lost, and Egypt was confined to the Nile Valley. By 959 BCE, the new state of Kush arose with its capital at Napata, and it eventually extended its control over southern Egypt. Circa 727 BCE, the Nubian king Painkhi conquered the rest of Egypt. Then in 664 BCE, the Assyrians invaded and sacked Thebes.

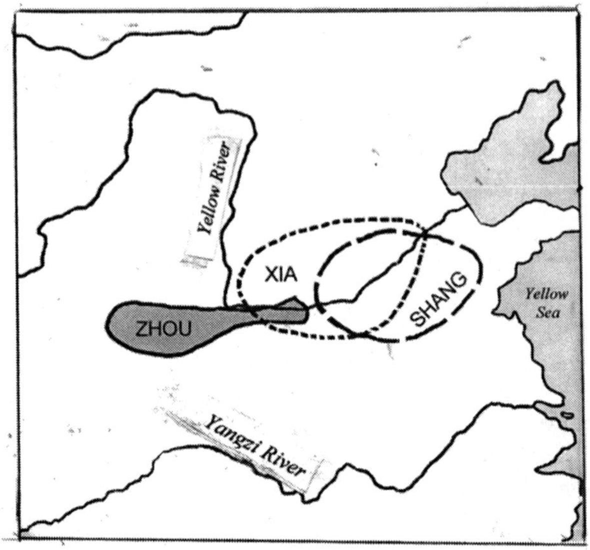

Figure 4.8. The Xia, Shang, and Zhou Dynasties circa 2000 BCE.

China. Far to the east, the Chinese began farming in the Yellow and Yangtze River Valleys about 8,000 years ago, about two millennium after agriculture first emerged in the Fertile Crescent. By the end of, the third millennium BCE, two states, the Xia the Shang had arisen in the Yellow River valley as shown in Figure 4.8 and became the first states with central governments in China. As in the west, their ruling dynasties had expanded their authority through warfare. In addition to warfare, both the Shang and Xia expanded their domains by building new towns in fertile areas along the Yellow River. Because of intensive agriculture, the populations of these two states grew until their territories met. Circa 180 0 BCE, the Shang defeated the Xia and claimed that they had received a "mandate of heaven" to rule the enlarged state.

In 1027 BCE, the Zhou Dynasty to the west challenged and defeated the Shang, united China, and claimed the "mandate of heaven" to rule all mankind for themselves. The Zhou expanded its territory in the north to the present border between China and Mongolia and then moved east and built a new capital at Xian, but over time, the control of the Zhou state gradually disintegrated into intrastate warfare, and in 771 BCE, the

Figure 4.9. Terracotta Soldiers of Emperor Zheng (photo by author).

western part of the Zhou domain was overrun by nomads from the north. The Zhou moved further east and founded a new capital at Louyang. Then other states that had previously acknowledged the Zhou's authority began to challenge it, and China's rival states then became embroiled in continuous warfare from 403 until 221 BCE, a period known as the Period of the Warring States. Each state maintained its own army of as many as 10,000 conscripted peasants, and the cities built walls for protection. In response, the armies developed siege engines to break down the walls. It was during this period that Sun Tzu wrote his famous *Art of War* at the end of the sixth century BCE. In addition to the interstate warfare, China was continuously threatened by nomadic Mongol tribes from the north. Between the fifth and third centuries BCE the Chinese states built walls along their borders which were the forerunner of the Great Wall of China to protect them from both the nomads and enemy states. The Warring States period lasted until 221 BCE, when the Quin state in the east achieved a decisive victory over the Zhou.[28]

During the fourth century BCE, the Qin state was organized along efficient, albeit ruthless lines and by 221 BCE, it had defeated all of its rivals and had unified China into a single state. Its ruler, Chang Yang (Lord Chang), assumed the title of the first emperor of China. Although

Chang's rule was harsh and repressive, he created an efficient bureaucracy, and is famous for his burial tomb, which contains more than 8,000 life size terracotta soldiers in battle formation with their horses and chariots as shown in Figure 4.9.

After his death in 338 BCE, Chang's kingdom was rapidly torn apart by the revolting cities, and by 108 BCE there were eighteen separate kingdoms in China. In 202 BCE, a new emperor, Liu Bang (Emperor Gsozu), emerged as the founder of the Han Dynasty, which survived more than 400 years, until 210 CE. The Han took over and improved upon the administrative system created by Lord Chang which consisted of a professional civil service system that evolved into China's mandarin system in which government officials are chosen by merit on the basis of a system of rigorous examinations. Using a conscripted peasant army, the Han extended the empire to include all of eastern China, the Korean Peninsula and the strip of land in central Asia that contained the Silk Road that extended to Ferghana in modern Uzbekistan.

The Han also extended the Great Wall and maintained garrisoned outposts along the wall to protect China from the Mongols. At its height under Emperor Wudi (r. 141–87 BCE), the Han empire included 57 million people, the largest state in the world at the time. The Han Dynasty collapsed in 220 CE when nomadic families that were permitted to settle inside the Great Wall revolted. The empire then split into the Three Kingdoms, an era that lasted until 265 CE, when, after a temporary unification (280–304 CE), there was a succession of six dynasties along and south of the Yangzi River, and North China dissolved into 16 warring kingdoms. China was not united again until the Sui Dynasty in 581 CE.

Although China was initially unified under the Zhou, China did not become a truly unified state until the Ming Dynasty in 1368CE. But even since that time China has been plagued with descent and rebellion. Today China faces descent in its Tibet, Hong Kong and Xinjing provinces and Taiwan which it claims is stilll part of mainland China.

The Western Hemisphere.

The first immigrants into the western hemisphere crossed the land bridge Beringia that existed between Siberia and Alaska during the last ice age, about 15,000 years ago.[29] The last migrations ended about 12,000 years ago, at the end of the ice age, when the land bridge disappeared because of the melting of the polar ice sheet that raised sea levels and engulfed Beringia, leaving only the Aleutian Islands. Over the following millennia, the immigrants spread out over the western hemisphere to the east coast of North America and to the southern tip of South America. North of the Rio Grande River, they mostly remained hunter-gatherers, except in the northeast. Farming began about 9,000 years ago in Mesoamerica, about two millennia after it did in the Middle East, Because of the poor soil and sparse rainfall in Mesoamerica, agriculture was slower to develop there than in South America and the settlers supplemented their diets by foraging and hunting.

In both Mesoamerica and South America, the Indians believed that man had been created from the blood of their gods, who demanded blood sacrifices from humans. Thus, human sacrificial rituals became the primary means for communicating with and placating their gods. For some of these cultures, war provided the primary means for obtaining sacrificial victims. Among the Aztecs, whose sacrifices were performed by cutting out the hearts of living victims, it was considered a great honor to be sacrificed, and there were many volunteers for the honor. Personal bloodletting was a way to enter into spiritual world and was the duty of kings and priests.[30]

The Aztecs. One of the tribes that settled in the Valley of Mexico—the present location of Mexico City—circa 1200 CE and established small citystates were the barbarous and uncouth Aztecs. At first they worked as laborers and mercenaries for groups that had already settled there, but in 1345 CE, they built their own city at Tenochtitlan on an island in Lake Texcoco. In 1427 CE, the Aztecs began a campaign of military conquests in which they were able to dominate most of Mesoamerica. At the beginning of their rise, the Aztecs formed alliances with the states of Texcoco and Tlacopan against Tlaxcala and other Mesoamerican states. In addition to fighting to extend their empire, the Aztecs also fought to obtain sacrificial victims. They believed that human blood was needed to make the sun rise each day, and daily sacrifices were made on the altars of the great temple at Tenochtitlan. They also believed that their battle fatalities insured the continuation of the world, and they referred to their wars as "flowery wars" because the flow of blood reminded them of falling flower blossoms.[31]

The Incas. Along the northwest coast of South America, the Andes Mountains provide a diverse ecology that resulted in several different lifestyles as different groups adapted to the local ecologies. The Pacific

Ocean provided a rich abundance of seafood for the coastal dwellers. When irrigated, cotton and other crops could be grown in the coastal valleys, and potatoes and cereals were cultivated on mountain terraces. At the highest elevations, llamas were herded as pack animals, and alpaca and vicuna were herded for their wool. This geographic diversity led to extensive trading in agricultural products between the regions.

South America was originally settled by migrants from the north who settled on the narrow coastal plain between the Andes Mountains and the Pacific Ocean, which provided a rich soil for growing maze and other crops and provided fish from the Pacific Ocean. The coasta l plain is inundated with seventy-eight short and narrow valleys of fertile land. The original settlers of these valleys became proto-agriculturalists, and over time improvements in agricultural resulted in growing populations, so that eventually all the available land in the valleys was brought under cultivation. Robert Carniero has postulated that when there still was a surplus of land, the agricultural communities engaged in primitive warfare. However as the land became scarce, the major incentive for war became to acquire territory from neighboring villages to feed the growing populations of the region.[39] Thus wars began to be fought for economic reasons, and the frequency, importance and intensity of warfare increased. Under these circumstances, a defeated village could either migrate into the highlands of the Andes, or be incorporated into t he victor's political system.

Those groups that chose to migrate into the Andes highlands settled in the Lambayeque Valley where they formed several small states, one of which was the Inca. Under King Pachacuti and his son, Topa, the Incas conquered all the other states in the valley and by 1527 CE they had created an empire that stretched 4,200 kilometers along the western coast of South America from Quito, Ecuador, to Valparaiso, Chile. They established their capital at Cuzco, which interprets as the naval of the earth.

The Incas ruled their empire by establishing administrative centers garrisoned by army troops throughout their empire. Their subjects were forced to pay heavy taxes in the form of a large fraction of their agricultural produce and in textiles and other craft products. They were also required to perform corvée labor to build a netwo rk of highways along the ridge of the Andes and through the coastal lands. Mountain chasms were spanned by rope bridges, and rest houses were built at intervals of a days journey by foot because the Incas did not have horses or other animals of burden.

Non-State Societies

In a few regions such as Greece, the Indus River Valley, and the Yucatan Peninsula, large states did not materialize, and in these regions the societies existed in separate citystates. The archaic Greeks, who had settled along the coasts of the Aegean Sea during the ninth century BCE, never developed a state but lived in fiercely independent citystates until Greece was unified by Philip II of Macedonia in 338 BCE. Disputes between the Greek citystates were, more often than not, settled by warfare, so that each citystate maintained its own army and navy, and warfare developed into a highly formal, ritualized affair conducted under an elaborate set of rules of hoplite warfare as discussed in the previous chapter.

Because most of the citystates were located on the coasts, each maintained its own navy and ships to transport its soldiers. Naval engagements were common, and the navies relied on oar power rather than on sails because of the greater maneuverability that oars provided. The early Greek naval ships were up to 20 meters long, with up to 50 oarsmen. By circa 500 BCE, the Greeks had developed the trireme, a fighting ship with up to 170 oarsmen seated in three tiers. The bow of the trireme was equipped with a ram that was the ship's main weapon.

Most of the wars between the citystates were over boundary disputes or failed alliances; few were fought for the conquest of territory. There was no sense of political unity among the citystates, and a spirit of intense rivalry existed among them. Thus Greek wars resembled those of primitive societies fought for personal stature and for sport, and were ends in themselves. When faced with a common threat, however, the citystates frequently formed alliances. At the beginning of the seventh century BCE, the Greek citystates united against Carthage, which was competing with the Greeks to establish colonies in Sicily and Italy. The Greek citystates united in a common cause, and both Italy and Sicily became battlegrounds between the two rival colonial powers. These conflicts continued until the later part of the fifth century BCE when Persia presented an existential threat to the Greek citystates. In 495 BCE, the Greek cities on the east coast of the Ægean Sea revolted against Persian rule and, with the help of the mainland Greek citystates, attacked and burned the Persian city of Sardis. The Persians retaliated by destroying many of the Ionian Greek citystates on the coast of the Ægean Sea. Then in 490 BCE, the Persian ruler, Darius, launched an invasion of Greece to punish Athens and Sparta for supporting the rebellion against Persian rule of the Greek Ægean cities . Darius landed an army of 20,000 men at Marathon but was defeated by an Athenian army of only 9,000 hoplites and was forced to withdraw. Then in 480 BCE, Xerxes, Darius's successor, launched a second invasion of Greece and defeated the Spartan army at Thermopylae in the mountains of central Greece. Xerxes then marched his army to Athens, whose population had been evacuated to the island of Salamis, killed its remaining defenders, and burned the city. The Persian fleet then proceeded to Salamis, where it met the Greek fleet and was routed with critical losses. The final battle occurred at Plataea, northwest of Athens, where the Greeks defeated the remnants of the Persian army. In the meantime, the Athenian navy found the Persian fleet beached at Mycale on the eastern coast of the Ægean and destroyed it. Following the Persian defeat, Athens was rebuilt, and the Golden Age of Athenian culture followed.[32]

After the defeat of the Persians, the rivalry among the Greek city-states returned, with Athens and Sparta competing to be the dominant power of the Mediterranean world. They had cooperated to defeat the Persians, and each city had its allies and supporters. Athens portrayed itself as the modern, liberal center of Greek culture and portrayed Sparta as reactionary and ruthless, but Athens was just as ruthless as Sparta. When the citizens of Lesbos rebelled against Athens in 427 BC E, the Athenian assembly voted to massacre all of the male inhabitants of Lesbos's capital, Mytilene. War finally broke out between Sparta and Athens in 461 BCE when Sparta attacked Athens. The Pelopon nesian War raged for 57 years, until Athens was depopulated by a plague a nd was forced to surrender to Sparta, thus ending Athens wer and influence and the golden age of Greek culture. After that, Greec e was dominated by a new power struggle between Sparta and Thebes, until Philip II of Macedonia won the Battle of Chaeronea in 338 BCE and then united the city-states under his leadership to form the first Panhellenic state.

Conclusion

In this chapter we have shown how war played a major role in the rise of civilizations. However the question arises of why nearly all of them followed the same pattern in their rise, why war played such an essential role in this process, and why this process occurred independently in so many places. These independent and mostly isolated events strongly suggest that this pattern is somehow encoded in the human genome.

Perhaps it is similar to the human linguistic ability. We don't inherit a specific language, but the general ability to use complex vocal symbols and to learn a specific language. The use of lethal force to create large societies may have arisen from the advantages that cooperative living provided humans, and the xenophobia humans have towards those who are not members of our tribe or clan. These questions are a challenge for future paleoanthropologists and human evolutionists.

Chapter 5

Ethnicity, Religion, Xenophobia and War

> Sympathy is directed solely, towards members of the same community, and therefore towards known, and more or less loved members but not all members of the same species.
>
> Charles Drwin[1]

> Arabs, Jews and Christians in Syria and Palestine, they are all alike, a beastly people. The whole lot of them is not worth a single Englishman!
>
> Sir Walter Norris Congrave[2]

> We cannot be their friends because they do not speak our language.
>
> an Afghan Soldier[3]

Wars can be divided into two major categories: interstate wars and non-state wars, which include insurgencies against an existing governments and ethnic and religious wars. Of the 28 ongoing wars listed in the September 2014 edition of Global Security's "The World at War," Sixteen were insurgencies, three were ethnic wars, three were religious wars and only three were interstate wars; three others could not be classified.[4] Of the 719 wars listed in the Correlates of War database that occurred between 1945 and 2008, 636 (88 percent) were non-state wars, mostly between ethnic groups, whereas only 83 (12 percent) were non-state wars. In this chapter we will investigate the causes of these non-state wars.[5] Firgure 5.1 shows how the number of non-state wars has increased since 1945, whereas the number of interstate wars rose slightly until about 1990 and then began to decrease during this period.

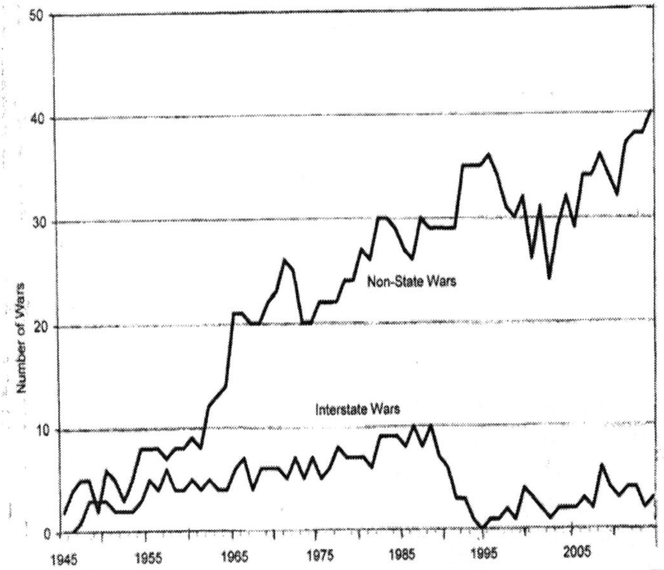

Figure 5.1 - Number of Non-state and Interstate Wars, 1945–2014.[6]

Non-State Wars

Non-state conflicts have accounted for the majority wars for many years, and many interstate wars were also ethnic or religious wars if the states involved consisted mostly of different ethnic or religious groups, such as wars between Christian and Muslim countries, or Sunni and Shiite Muslims countries, or ethnic groups such as the Dinka and Nuer ethnic groups in South Sudan.

Ethnic and religious wars are usually driven by the absolute values that their contestants are unwilling to compromise. Compared to international wars, ethnic and religious wars are especially costly in terms of human life because of the extreme xenophobia that exists between the belligerents, and they usually last longer than interstate wars. The average length of non-state wars that began between 1970 and 1999 was about six years, whereas the average length of interstate wars that began during the same period was only about two years. Some of the non-state wars have lasted for decades, such as the Northern Ireland Troubles, the Angola Civil War and the Israeli-Palestinian conflict. The objective of most interstate wars is to defeat the opposing states army, either in the field or by destroying its ability to equip and supply its army, or by both. In most non-state wars, especially in ethnic and religious wars, the objective is often to destroy the enemy's entire population. D.A. Summerfield has estimated that civilians account for over 80 percent of the fatalities in current non-state wars.[7] whereas during World War II, for example, civilian casualties were about 52 percent of all the casualties.

Following the defeat of the Iraqi army in May 2003, the Iraqi War devolved into a combined religious and ethnic war involving the Shiite and Sunni Muslim sects, the Iraqi Kurds. and the occupying coalition force that was attempting to establish a secular, democratic coalition government in Iraq. Following the withdrawal of the coalition forces in December 2011, the three ethnic groups continued fighting each other, the Sunnis and Shiites for control of the Iraqi government, and the Kurds for their independence.

Ethnicity

Ethnicity has been the cause of many of the contemporary wars. An ethnic group is a group of humans who perceive themselves to be a unique group, such as a race, a nationality, a group that shares a common language, religion, homeland, cultural heritage, or ancestry, and that may have similar physical traits such as skin color, stature, hair or other physical features. These shared characteristics and beliefs have powerful emotional and psychological attractions that result in the development of strong affiliative bonds between the members of the group and toward the group as a whole. Such bonds form the basis of an ethnic group that divides the world into "us" and "them," and such groups nearly always become ethnocentric and see themselves as unique and superior to other such groups.

A significant characteristic of an ethnic group is that it follows a double standard of morality: one kind toward its fellow members, and another toward all others. The Golden Rule, "Do unto others as you would have them do unto you," many times does not extend beyond ones ethnic group. Although the second of the Ten Commandments, "Thou shall not kill," condemns murder, the same book of the Bible sanctions war against and the killing of the enemies of the Jews: "The Lord is a warrior. . . . The chariots of the Pharaoh and his army, he has cast into the sea; the flower of his officers are engulfed in the Red Sea s. The watery abyss has covered them, they sank into the depths like a stone. Thy right hand, O Lord, is majestic in strength, thy right hand, O Lord, shattered the enemy."8 In the book of Deuteronomy, God commands the Jews to conquer the land of Canaan and instructs them on how to fight,

> When you take to the field against an enemy and are faced by horse and chariots and an enemy greater than yours do not be afraid of them, for the Lord your God who brought you out of Egypt, will be with you. . . . In the cities of these nations whose land the Lord your God is giving you as a patrimony, you shall not leave any creature alive. You shall annihilate them—Hittites, Amorites, Canaanites, Perizzites, Hivites, Jebusites—as the Lord your God commanded you.[9]

Muslims divide the world into *dãr al-Islam*—the house of submission, or the land of Islam—which is in perpetual conflict with *dãr al-Harb*—the land sanctioned for war where the infidels (nonbelievers) live, who must either convert to Islam or be killed.[10] Some Muslims believe that the revelations that Muhammad received from the Archangel Gabriel were meant for all mankind and that it is the Muslims duty to convert those to Islam who have not accepted its revelations.

A certain degree of ethnocentrism is probably essential if a culture is to persist. As children grow, they are taught the values of their own culture and that these values are superior to those of other cultures. The empathy that humans feel toward the other members of their own group and toward the group as a whole, and their willingness to risk their lives for their group or country is known as ingroup altruism or parochial altruism[11] as opposed to humanistic or universal altruism, which is directed toward all humanity, not only to members of one's own group. Parochial altruism is a form of behavior in which an individual assumes risks for the benefit of his group. Charles Darwin was one of the first to recognize that altruism and human sympathy may not extend beyond the members of one's tribe or ethnic group, as expressed in the epigraph at the beginning of this chapter and in the following, "But they [social virtues] are practiced almost exclusively in relation to men of the same tribe; and their opposites are not regarded as crimes in relation to the men of other tribes."[12]

Ethnicity is easily recognized. During a pub tour of London's East End by Senator Daniel Moynihan and his Scotch-American companion in 1966, they accepted an invitation to try their hands at a game of darts. After two or three minutes an English lady who was sitting on a nearby bench remarked, "E';s Irish, 'e's Scotch, both Ni'vy

(Navy)," which was true.[13]

The size of an ethnic group can range from a few thousand people, such as the !Kung who live in the bush on the edge of the Kalahari Desert in Namibia and Botswana, to hundreds of millions of people of a modern nation-state. Thus, the French, British, Germans, Italians, Greeks, Spaniards, Iranians, Han Chinese, and Japanese can each be considered to be a unique ethnic group because they all differentiate themselves from other ethnic groups by their language, dress, and/or culture. On the other hand, a state may contain many ethnic groups, whose members may also consider themselves to be united into a single ethnic group. Although the United States is made up of Anglo-Americans, African-Americans, Latinos, Irish-Americans, Italian-Americans, Chinese-Americans, and other similar ethnic groups, many residents of the United States consider themselves first to be ethnic Americans. Such ethnic groups are often formed through the intermarriage of persons of different races or ethnicity, in employment situations or in mixed ethnic or racial social groups.

Some of the largest ethnic groups are nations or civilizations which are usually characterized by a common language, history, culture, homeland, and/or religion.[14] Thus, we can speak of the German, Japanese, or Russian nations. Most nations or civilizations that exist today are the result of the successful spread of the culture of a smaller ethnic group, either through conquest or by migration. Most civilizations, including the Western, Chinese, Russian, and Islamic civilizations, were spread mainly through military conquest, and most are united by a common religion.

The major civilizations that exist today include: Western Civilization, which extends over Western Europe from Poland's eastern border, the western hemisphere, Australia and New Zeeland, where the predominant religions are Roman Catholicism, its Protestant offshoots and Judaism; the Russian civilization in which the Russian Orthodox religion is the dominant religion; Chinese or Sinic civilization in which Confucianism and Buddhism are the dominant religions; Muslim civilization is characterized by Islam; Hinduism is the dominant religion of the Indic civilization.; and Persian civilization is characterized by Zoroastrianism and Farsi as the common language, although almost all Iranians are now Muslims. Both the Sinic and Muslim civilizations use common written languages, Mandarin and Arabic respectively, although there are many spoken dialec ts in both civilizations. The Sinic civilization has spawned a number of secondary civilizations that include the Japanese, Korean, Tibetan, and Vietnamese cultures or sub-civilizations. History has seen major wars between civilizations, of which the Crusades between Western and Islamic civilizations in the Middle Ages, and the Cold War between the Russia and Western civilizations are examples.

Ethnic groups are seldom stable. Some are constantly disappearing, usually through the merger with other ethnic groups or by intermarriage, and new ones sometimes form, usually because of the fission of one ethnic group into two or more new ones. Until the Roman invasion, the population of England consisted mostly of Celts who immigrated into the British Isles during the Iron Age—the eighth to first centuries BCE. The Saxons and other Germanic tribes migrated into the Isles in the fifth century CE, and the English nation began to form when the occupants united against the invasion of the Danish Vikings in the ninth century CE. After the Norman invasion in 1066 CE, the Normans assimilated themselves into the English culture and adopted the English language although they introduced some French words into the English language.

Prior to the fifteenth century CE, the Plains Indians of North America constituted a single tribe from which the Osage, Kansa, Omaha, Ponca, Quapaw, and other Plains tribes broke away one by one to form separate ethnic groups that fought each other. The Dutch immigrants who settled in the Cape region of Southern Africa during the latter half of the 17th century developed their own distinctive identity, culture, and a dialect of Dutch, Afrikaans, and they no longer consider themselves to be Dutch.

In some cases, the migrants into a region retain their ethnic identities, such as the Catholic Croats, the Orthodox Serbs, and the Muslim Turks did after they migrated into Balkan Peninsula, and as the Tutsis did when they migrated from eastern Africa into Rwanda and Burundi. Sometime in the future, Europeans of the European Union may become united into a single ethnic group and consider themselves to be ethnic Europeans.

After the Manchu conquest of China in 1644, their overthrow of Ming Dynasty and their establishment of the Qing Dynasty, the invaders assimilated themselves into the Chinese culture, adopted. Confucianism and Chinese Buddhism, and used them to support their own political authority. They promoted the study of the Chinese classics and the veneration of ancestors. They became ethnic Chinese and even "more Chinese" than the indigenous Han Chinese.[15]

People are frequently able to voluntarily change their ethnic affiliation. In Rwanda, two ethnic groups, the Tutsis and Hutus, constitute the majority of the population. The Hutus are principally farmers, whereas the Tutsis are mostly herders and are believed to have immigrated later into Rwanda from the northeast. The original Tutsis were lighter-skinned and taller than the original Hutus, however intermarriage over many generations has eliminated the morphological differences between the two groups, and today, ethnici ty is determined mostly by occupation and to a lesser extent by ancestry. If a Hutu farmer receives a cow from a Tutsi, which by tradition, he is not supposed to possess because it is a sign of wealth and good breeding, it would be a path to upward social mobility and he would be de-Hutuized and become a Tutsi. Similarly, if a Tutsi herder lost all of his cattle and took up farming, he would become a Hutu.[16]

Xenophobia

In Chapter 3, we described how warfare was endemic among primitive human groups. Hostility toward outsiders who are not members of ones group is manifested as xenophobia, which is the intolerance, fear, and contempt that one group of social animals has toward another group of its conspecifics. In humans xenophobia is frequently the cause of war. According to the sociobiologist Edward O. Wilson, "The strongest evoker of aggressive response in animals is the sight of a stranger, especially a territorial invader."[17] Xenophobia has been documented in virtually every species of social animals, including ants, chickens, Canada geese, chimpanzees, and rhesus monkeys.[18] Some biologists believe that xenophobia evolved among social animals from conflicts over the need to acquire and defend territory, food, mates, and offspring of their social group. One of the features of xenophobia is the ability of the members of one ethnic group to distinguish themselves from members of another group—the ability to distinguish "us" from "them." In the summer of 1858, Charles Darwin placed a few red ants close to another nest of red ants, and he wrote in his diary, "I pass my time c hiefly in watching ants & find that though many thousands inhabit each hillock, each seems to know all its comrades, for they pitch unmercifully into a stranger brought from another ant hill."[19] Insects are able to distinguish members of their own colony from those of another colony by the pheromones (odors) they emit, just as human armies differentiate themselves by their uniforms. Agonistic behavior between animal groups can range from competition over food and territory, and ritualized displays and chasing, to lethal combat. According to Jared Diamond, "Xenophobia comes especially naturally to our species, because so much of our behavior is culturally rather than genetically specified, and because cultural differences among human populations are so marked. Those features make it easy for us, unlike wolves and chimps, to recognize members of other groups at a glance by their clothes and hair style."[20]

Humans, like most other animals, tend to be distrustful of strangers. Human infants are instinctively afraid of strangers and cry when confronted with the face of a stranger; however after they become accustomed to the strange face, their fear ceases. Social introductions to strangers probably originated to overcome this innate fear of strangers. A significant driving force behind xenophobia may be the high cost of not being suspicious of strangers and of being wrong, so that our suspicion and distrust of strangers and outsiders is likely to be a product of our evolution and is embedded in our genome.[21] Many ethnic groups develop strong xenophobic feelings toward other groups, especially those that occupy the same territory, such as Catholics and Protestants in Northern Ireland; Catholics, Orthodox Christians, and Muslims in the Balkan Peninsula; Jews and Palestinians in the Levant; Hindus and Muslims in Kashmir; Christians and Muslims in sub-Saharan Africa; and black farmers and Muslim herders in the Darfur region of Sudan.

The island of Hispaniola is occupied by the Dominican Republic on the east end and by Haiti on the west end of the island. Although the inhabitants of both countries trace their ancestry to the African slaves brought to the island in the sixteenth and seventeeth centuries, the skin color of the Haitians is generally darker than that of the Dominicans. The Dominican Republic's constitution grants Dominican citizenship to any person born in the country, but in 2013 the Dominican Supreme Court ruled that Haitian workers were considered to be "in transit" and that their children born in the Dominican Republic were not entitled to Dominican citizenship. The ruling rendered tens of thousands of Haitians born in the Dominican Republic stateless. Responding to international pressure, in 2014 the Dominican government established a process for obtaining citizenship that required Haitians born in the country to provide either a valid birth certificate which in most cases was nor issued, or notarized letters which are very difficult to obtain and which human rights advocates said sets too high a bar. The US State Department told the New York Times that it was "concerned that eligible individuals may not have had sufficient time and means" to have their applications evaluated before the Dominican government stopped accepting applications for citizenship in February 2015, leaving tens of thousands of people stateless. [22]

On a single day in August 2012, the *New York Times* ran three stories regarding ethnic xenophobia. In the first story, the governor of Russia's Krasnodar region, that hosted the 2014 Winter Olympics, enlisted the area's Cossacks as an auxiliary police force to prevent dark-skinned Muslims from the North Caucasus from moving into his region. He said, "Who will answer when the first blood is spilled, when interethnic conflicts start? And sooner or later it will happen." He offered Kosovo as an example, saying that "Albanian [Muslims] began to destroy churches, forced the dominance of their culture, their religion, began conflicts, imposed pressure, blood, small war, big war. And that was it, there was no country, there were no people, thousands of refugees all over the world."[23]

In the second *New York Times* story, during the month of Ramadan, local governments in the western region of China's Xinjiang Province ordered Muslim restaurants to stay open and told civil servants and students to continue to eat during daylight hours, when Muslims usually fast during Ramadan. A notice on the Web site of the health bureau of Urumqi, the regional capital, said local officials had discussed "increasing eating and drinking during Ramadan" and had been urged "to increase the monitoring of collective eating with supplied food at work units." The goal was to "guarantee the health of the masses."[24] China has been promoting the immigration of Han Chinese into southern Xinjiang, which has a large Muslim Uighur population.

Finally, the *New York Times* reported that in a Pashtun village in Afghanistan's eastern province of Oruzgan, at least nine people were forced from their homes and executed by a Hazara commander, who said that the executions were to avenge the killing of two Hazara men by the Taliban the previous day. Most Taliban are ethnic Pashtuns and Sunni Muslims, whereas the Hazaras are Shiite Muslims. Around 200 Pashtuns from neighboring villages where the executions occurred congregated in the center of the district, shouting for justice. Khas Oruzgan is a mixed Pashtun and Hazara area where tensions between the two groups are decades old. When the Taliban governed the country in the 1990s, the Hazaras were victims of Taliban persecution.[25]

Sectarian xenophobia has been particularly prevalent in those Muslim countries which are ruled by minority sects, such as in Syria

Bahrain, and Iraq before the US led coalition invaded Iraq in 2003, and has been the cause of uprisings of the majority sects in Syria and Bahrain against the ruling minority government during the 2011–2012 Arab Spring, and in Iraq after the defeat of the Iraqi Army to the invading coalition force.

Genocide has become the most violent and extreme manifestation of xenophobia. In many cultures that incorporate several ethnicities and have had a minimum of ethnic and racial conflicts in the past, ethnic conflict can suddenly break out, as they did in the former Yugoslavia in 1993, in Rwanda in 1994, in the Darfur region of the Sudan in the 1990s, and in Kenya after the presidential elections in December 2007 and March 2013. Until 2007 Kenya had been regarded as one of the most stable countries in Africa and had been the envy of many for its stability and prosperity, yet after the most fiercely contested elections in the country's history, in which Kenya's election commission ignored undeniable rigging of the election by the country's ruling party, rioting broke out between the country's two major ethnic groups, the Luos and the ruling Kykuyus. The election seemed to have tapped into an atavistic strain of ethnic tension between the groups that had lain just below the surface but until the election had not provoked widespread mayhem.[26]

Slavery is another form of xenophobia, and in ancient times, it manifested itself in the enslavement of war prisoners. Nearly all ancient societies from Sumer to Rome engaged in this practice, and slaves were usually treated as non- or sub-human.

The relationships between human ethnic and racial groups are often ambiguous and unstable, and can change rapidly from cooperation and hostility. Different ethnic and racial groups often live together in the same communities for long periods of time, working together and intermarrying, and then some event triggers the outbreak of intergrou p violence, such as the murder of Martin Luther King or of the killing of a black teenager by a white police officer that sparked race riots in the United States. Another example is the ethnic cleansing in Bosnia-Herzegovina during the 1990s when over 10,000 Bosnian Muslims were murdered by Orthodox Christian Serbs. The ethnic cleaning was triggered by the breakup of the former Yugoslavia and the formation of the Republic of Bosnia-Herzegovina and the Serbian Republika Srpska. Before the outbreak of the civil war in Bosnia in 1992 the populations of Bosnia-Herzegovania consisted of 44 percent Muslims, 32.5 percent Serbs and 17 percent Croats.

Religion and War

Previously in this chapter, we saw how the Jewish God commanded his people to destroy their enemies and to conquer their lands to expand the territory of Israel. They were to show no mercy to women, children, or even to babes in arms. Thus the god of the ancient Jews sanctioned wars of conquest.

In Islam, Mohammad taught that *jihad*—holy war—is the sacred duty of Muslims to spread Islam to all men and to kill the infidels: "Then, when the sacred months* are over, kill the idolaters wherever you find them, take them [as captives], besiege them, and lie in wait for them at every point of observation. If they repent afterwards, perform the prayer and pay the alms, then release them. Allah is truly All-Forgiving, Merciful."[29] Regarding Jews and Christians, "Fight those among the People of the Book (Jews and Christians) who do not believe in Allah and the Last Day, [who] do not forbid what Allah and his Apostle have forbidden and [who] do not profess the true religion, till they pay the poll tax out of hand and submissively."[30]

Hindu attitudes toward war are addressed in the *Bhagavad Gita*. When the armies of the *Kauravas* and the *Pandavas* were arrayed against each other waiting for the signal to begin fighting, Arjuna, the leader of the *Pandava*, upon seeing his relatives and friends arrayed against each other, was suddenly overcome with a deep anxiety. He felt that it would be a sin to kill his kinsmen for the sake of his. kingdom, so he laid down his bow and told his friend Krishna that he could not fight. Krishna responded that Arjuna would commit a sin if he failed to perform his duty as a warrior and that he must fight. Krishna reminded Arjuna that life is eternal, that there was never a time when neither they nor the kings arrayed against them had not existed in the eternal cycle of life.[31]

In China, Confucius and his disciple Mencius taught that states, in dealing with one another, should observe good faith and moderation in war, avoid imperialism and intervention in the affairs of other states, and avoid aggressive wars of conquest. Confucius was not a pacifist;. although he regarded war as evil, he said that when it comes, it must be waged vigorously.

Christian attitudes toward war are embodied in the just war theory of Augustine of Hippo (St. Augustine), who, in the fourth century CE

* The three months in which inter-tribal fighting was prohibited in Arabia

formulated standards for going to war—*ius ad bellum*—and for its conduct—*ius in bello*. Augustine's just war theory argues that a state formulated standards for going to war—*ius ad bellum*—and for its may have to resort to war as an instrument of justice for the common good of mankind but that rulers are enjoined against resorting to war unless they are morally certain that their cause is just and that their rights have been violated. There also must be a reasonable prospect that the use of force will produce more good than evil and that it will restore justice.

In Augustine's just war tradition, *ius in bello* requires that military actions be discriminate in the sense that the military action is appropriate to the situation and avoids attacking noncombatants or causing unnecessary killing of civilians or the destruction of nonmilitary installations, even if unintentional. They must be proportionate in the sense that the military action is appropriate to the situation and avoids unnecessary use of force; and they must be militarily necessary in the sense that military actions are limited only to those necessary to achieve legitimate military aims.

However these criteria of the just war theory are subjective and can be used to justify almost any aggression or military action. Augustine formulated his just war theory as a response to the early Christians' pacifism so that Christians could fight on behalf of Rome to enforce its imperial peace, *pax Romana*, but they had to fight justly for that peace.[32] Most just war theorists seem to regard war as a necessary evil that needs to be justified; however the 15th-century Dutch philosopher Erasmus maintained that there was no such thing as a just war.[33]

Like most other religions, Christianity does not condemn war but has sanctioned it to achieve its own goals. In response to an appeal by the Byzantine emperor Alexius for aid in turning back the "Islamic tide" that threatened Constantinople and the Byzantine Empire in 1095, Pope Urban II appealed to the knights in Western Europe to undertake a crusade and a holy war to liberate Jerusalem from the Muslims. His battle cry was "God wills it!" He initiated two centuries of futile warfare that left Palestine in the hands of the Muslims. The religious wars of the 16th and 17th centuries between Catholics and Protestants in Northern Europe were justified on religious grounds, but an important cause of these conflicts was the desire of the Protestant rulers in northern Germany to escape papal domination. Although the conquests of Spain and Portugal in the western hemisphere were justified on religious grounds to convert the native Indians to

Christianity, their true purpose was to dominate the Americas and to enrich Spain's royal coffers with American gold and silver.

Nationalism and Patriotism

In most cultures governance and religion are intertwined and the head of state is also the religious leader. During the reformation in Europe in the sixteenth and seventeenth centuries, the rulers would often replace sacred art with paintings of themselves in the churches and require that their subjects worship them. In Russia the Tsar was known as the Little Father and in China the emperor had the Mandate of Heaven to rule over all mankind. Traditionalist Muslims have no conception of the separation of mosque and state and the Caliph is both the religious and political leader. Thus religious fidelity and patriotism were identical.

However, in western countries in which there is a separation between church and state; in communist countries which forbid theistic religions; and in secular Muslim countries that are usually military dictatorships, a national civil religion develops that unites the citizens of the state into a nation, and the resulting set of beliefs and loyalties to them is called patriotism. A civil religion may include such things as the belief in and practice of representative democracy, freedom of speech and the press, human rights, and freedom of religion, or it may be the allegiance to the government, a monarch or the head of state. In some states, such as the former Soviet Union, China, and North Korea, theistic religion has been replaced by state religions that worships the state and its rulers. In monarchies such as Japan, the king or emperor often assumes the role of the common ancestor. In other countries, a national hero, such as a founding father, assumes that role. In the United States, the Pilgrim Fathers, George Washington, Thomas Jefferson, and Abraham Lincoln are the common ancestors. Phenotypic traits that distinguish some ethnic groups from others can be replaced by common apparel worn by most members of the nation.

Japan is an archetypical example of the nation-state. All Japanese speak the same language, are morphologically similar, and share the same homeland, and most practice the national religions of Shintoism and Mahayana Buddhism. All of these features are fused together in the person of the emperor, who personifies the divine origin of the Japanese nation and is the "father" of the nation. Akihito, the current emperor, traces his ancestry back 70 generations to the first emperor, Jimmu, who founded the empire in 660 BCE, and symbolizes the common origin of the Japanese people. Until Japan's defeat in World War II, the institution of the emperor had become a cult that permeated every aspect of Japanese society. The Japanese people had immortalized the emperor and believed that they were his children. As a result of their geographic isolation, their common origin, and their deification of the emperor, the Japanese are one of the most ethnically and culturally homogenous nations today.[27]

The Genetic Basis of Ethnic Xenophobia

Several recent studies have shown that there is a genetic basis for racism and ethnic xenophobia. One by Shihui Han and his colleagues at Peking University in China has shown that there is a neurological basis for racial xenophobia.[28] Using functional magnetic resonance imaging (fMRI), they observed the responses in the brains of Caucasian and Chinese subjects while they watched video clips of the faces of Caucasian and Chinese individuals receiving painful and non-painful stimulations. They found that when a subject was shown clips of individuals of the same race as the subject receiving painful stimulations, activity in the anterior cingulate cortex (ACC) region of the subject's brain increased. The ACC responds to conflict within an individual. But when a subject was shown clips of individuals of the other race receiving painful stimuli, activity in their ACC decreased significantly. This was true of both Chinese and Caucasian subjects, strongly suggesting that there is a neurological basis for racial xenophobia.

There is a rare genetic disorder known as Williams syndrome. Children who are afflicted with the malady display extreme sociality in addition to having mild to moderate learning difficulties. Unlike normal children, they lack the fear of strange or angry faces and display a lack of racial biases. The condition appears to be caused by reduced neural activity in the amygdalae—the region of the brain that responds to threatening situations that result in fear. The condition is recessive and has been attributed to the deletion of 28 genes on the seventh chromosome, which implies that intergroup xenophobia is due to these specific genes.[29,30,31] These genes evidently mediate a person's response to strangers to make them suspicious of strangers.

However in spite of the human predisposition towards racism and ethnic xenophobia, there are many instances in which such prejudices have been overcome and culturally suppressed, and people of different races and ethnicities have learned to peacefully live together and intermarry, but then some event such as the assassination of a revered leader, or the killing of a man of color by a white police officer activates our genetic predispositions towards racial or ethnic xenophobia that had been culturally suppressed by "the better angles of our nature." Thus our genetic predispositions towards racism and ethnic xenophobia have condemned us to our baser instincts—the devil within us—that prevents us from achieving the just societies that many of us dream of.

Chapter 6

Why Humans Make War

If I die mother, don't cry for me, I'll have died so my country can live.

Mohi Salah[1]

I thank God for this gift. This is the tenth member of my family to receive the honor of martyrdom.

Khalil al-Hayya[2]

If you are going to go to war or start an insurrection, you will need an army of soldiers or suicide bombers who are willing to fight and kill other human beings, and to risk their lives doing so. It turns out that it is easy to find such groups of human beings. Take any group of boys or young men between the ages of 10 and 25, put them through a course of basic training, and imbue them with the virtues of their nation, religion, or political ideology, and *voila*: you have an army, a group of suicide bombers, or a squadron of kamikaze pilots eager to kill other human beings and willing to sacrifice their own lives doing so. Unless warriors are willing to risk their lives in aggressive behavior, it is unlikely that human warfare would ever have evolved. In this chapter we will explore why humans are so willing to fight and die for what they believe in.

Willingness to Die

The will to survive is an inherited instinct so that one can mate and pass one's genes on to her or his offspring, yet many humans are willing to forgo this opportunity and willingly risk their lives in a variety of situations. Explorers, adventures, and other thrill seekers knowingly risk their lives to achieve their goals. All the members of Ernest Shakelton's two Antarctic expeditions, 1907–1909 and 1914–1916, were fully aware of the life threatening risks they were taking when they agreed to accompany Shakelton, yet their eagerness for the adventure was greater than their fear of death. Mountaineers who climb Mount Everest and other lofty peaks are also fully aware of the risks they take. Yet every year, several hundred climbers risk their lives to reach the summit of Mount Everest. In over 2,200 attempts since 1922 there have been 185 deaths, a fatality rate of 8.4 percent, similar to that of many wars.

A distinctive feature of humans, primarily young males, is their willingness to take existential risks for their country, ethnic group, or religion; and such groups and organizations willingly sacrifice their young men in war for the group's survival and aggrandizement. In many wars, men willingly go into battle knowing that there is a high probability that they may be killed or severely wounded. Some of the worst of these battles occurred during the First World War. In the First Battle of the Marne in September 1914, four German armies containing about 1.275 million men attacked six French armies and the British Expeditionary Force, a total of 1.125 million men, in an attempt to capture Paris. During the battle, the French and British forces suffered 263,000 causalities (dead and wounded), or 24 percent of their forces, and the Germans suffered 220,000 causalities, or 17 percent.[3]

In the 1916 Battle of the Somme, which lasted from July 1 to November 13, the Allied forces attempted to drive the German Army from France. By the end of the battle, the French and British had committed 99 divisions containing about 1.2 million men to the battle, and the Germans had committed 50 divisions containing about 750,000 soldiers. On the first day of the battle, the British suffered 60,000 causalities—20,000 dead and 40,000 wounded—the most causalities ever suffered by the British Army in a single day, and by the end of the battle, the Allied forces suffered 614,100 causalities, or 51 percent, and the Germans suffered more than 600,000 causalities.[4] During the four-month battle, the Allies advanced only about 11 kilometers along a 20 kilometer front—a very high price to pay for such a small gain.[5] More than 16 million people lost their lives during World War I, and it is said that Europe lost a generation of its manhood during the war. During the American Civil War, 365,000 (17 percent) of the Union soldiers and 260,000 (24 percent) Confederate soldiers lost their lives, some of the highest fatality rates of any war ever fought. Why are men so willing to engage in such battles, and why are their leaders so willing to commit these men to almost certain death? In one sense, these soldiers are like the workers in an anthill or a beehive when they risk their reproductive abilities for the benefit of their colony.

Why Men Are Willing to Fight and Die for Their Cause

An example of how soldiers can be motivated to support the aggressive goals of their national leaders are provided by the comments of a member of the 24th German Panzer Division during World War II: "Most of us . . . believed in the righteousness of our cause. . . . I never could say about myself that I was an ardent Nazi. I didn't know enough about it. But I believe that it was my right to enforce our way onto others and for that reason I was doing a God given job to go into other countries, and how can you do it better than with a Panzer."[6]

Some religions promise rewards in the afterlife to those who die defending their faith. An example is the account of the heroic death of Jaklin, a knight of the order of the Knights Templar from Touraine, France, who was killed fighting a Muslim raiding party in Galilee in May 1187 during the Third Crusade. He died surrounded and hopelessly outnumbered. His death was recorded thus,

> He was not afraid to die for Christ. At long last, crushed rather than conquered by spears, stones and lances, he sank to the ground and joyfully passed to heaven with the martyr's crown, triumphant. It was indeed a gentle death with no place for sorrow, when one man's sword had constructed such a great crown for himself from the crowd laid all about him. Death is sweet when the victor lies encircled by the impious people he has slain with his victorious right hand. . . . Such a great number of Turks had rushed in to attack, and this one man had fought for so long against so many battalions, that the field in which they stood was completely reduced to dust. . . . It is said that there were some who sprinkled the body of the dead man with dust and placed dust on their heads, believing that they would draw courage from the contact. In fact, rumor has it that one persson was moved with more fervor than the rest. He

> cut off the man's genitals, and kept them safe for begetting children so that even when dead the man's memb ers—if such a thing were possible—would produce an heir with courage as great as his.[7]

The soldier had died a proud and peaceful death, satisfied that he had fulfilled his mission and had served his god.

Islam promises its martyrs rewards in paradise to those who die for Allah, and martyrs are promised companionship with Allah and the Prophet Mohammad, and holy virgins when they reach paradise. During the Syrian Civil War, jihadist fighters from Jordon crossed the border into Syria to join the Nusra Front, an Al Qaeda affiliate, or the Islamic State, often leaving their families to fight in Syria. These men were driven by an intense conviction to serve Allah, and they saw Syria as a launching pad for founding an Islamic caliphate governed by Sharia law. Abu Abdullah abandoned his wife and two children to join the Nusra Front. He relished the feeling of serving in what he felt was a celestial cause to found an Islamic Caliphate, but his wife's anguish persuaded him to return home, where he longed to return to the war: "If I could go back and do it again, I would not come back. Those were the best three months of my life." Mohammed Abu Rahaim's two sons had joined the Nusra Front in Syria. One, Harith, had left his wife and three children to fight in Syria. Mr. Rahaim showed a picture of Harith with a rifle at his side, heating water over an open fire. The next photo showed Harith dead with his bearded face protruding from a body bag. When Mr. Rahaim received the news of his son's death while driving, he said, "I got out of my car and bowed to God." He said that he was proud that his son had achieved the martyrdom that he so desired.[8]

One fighter for the Islamic State, who had shot two enemy men and had participated in a public execution, said, "When you fight over there, it's like being in a trance. Everyone shouts, 'Allah Akbar (God is great),' which gives you divine strength to kill the enemy without being fazed by blood or splattered guts." Can, his middle name, was a drug addict from one of Ankara's poorest neighborhoods. He said that he didn't think that he had much to loose when he was smuggled into Syria. Many of the Turkish fighters in the Islamic State say that they identify more with the extreme form of Islamic governance practiced by the Islamic State than with the more moderate Islamic rule practiced by Turkey's governing party.[9]

Strange as it may seem to those who have never been in combat, many soldiers find great pleasure in killing their enemy. Henrydel Mann, a Belgian soldier, wrote of his experiences during World War I,

> I had thought of myself more or less immune from this intoxication, until as a trench mortar officer . . . I secured a direct hit on an enemy encampment, saw bodies and parts of bodies go up in the air, and heard the desperate yelling of the wounded or runaways. I had to confess to myself that it was one of the happiest moments of my life.[10]

Flight Lt. D.M. Cook, a British Spitfire pilot in World War II, described the moments before an aerial dogfight as "the most gloriously exciting moment of my life;" he was absolutely fascinated by the sight of an enemy plane going down.[11] Roderick Chisholm, a night fighter pilot in the Royal Air Force, described shooting down two German planes in one night as "sweet and exciting."[12]

When he was president of the United States, Theodore Roosevelt claimed that his charge up San Juan Hill in Cuba during the Spanish-American War "was the great day of [my] life."[13] General Robert E. Lee once remarked, "It is well that war is so terrible or we would grow too fond of it."[14] The pleasure and thrill that a soldier experiences in killing his enemy is surely an evolutionary adaption to take existential risks for the survival and aggrandizement of his country, religion, or ethnic group.

Love of Comrades and Small Unit Cohesion. In addition to the pleasures that men find in killing their enemies, and the deeply felt emotions that most men feel for their country or religion, a key motivation for fighting is the bond that develops among men who fight together in small units of five to ten soldiers—an infantry squad, the crew of a mortar or artillery piece, a tank, or an aircraft. Small unit cohesion depends on the ability of a unit's members to form close affiliative bonds, and one of the major objectives of basic training in all armies is to create this group bonding and cohesion. In the US Army, new companies are formed from recruits who train together and remain together as a unit throughout their military careers, so that close cohesive bonds are formed between the members of the unit.[15] In the US Army, replacements for causalities are usually soldiers who have just completed their basic training and who have not yet been in combat, so they easily bond with the other members of the unit to which they are first assigned. In the German Army, combat units are formed from men from the same village or neighborhood of larger cities that are likely to have known each other since childhood or to have common points of reference, so they will easily bond together and form a cohesive unit. Such units have proven to be highly effective in combat, and there is an intense loyalty among the members of such units. The commanding officer of a US Marine platoon fighting against insurgents in Iraq wrote, "Day after unrelenting day bound our platoon tightly together, eventually creating a whole much greater than the sum of its parts, and we grew to love one another fiercely. I knew these men better than my best friends; better, in some ways, than my wife."[16]

Combat units become surrogate families for their members. According to one combat veteran, "Among men who fight together, there is an intense love. You are closer to these men, except for your immediate family when you were young."[17] "In battle, the unit becomes the only important thing in the infantryman's universe, nothing outside it matters, and no sacrifice for the other men in it is too great."[18] A 14th-century knight said, "You love your comrade so much in war. . . . and you are prepared to go and die or live with him . . . [and are] afraid of nothing."[19]

William Manchester, the American historian and biographer who left the safety of the hospital to return to his unit in the Pacific theater during World War II, wrote,

> Those men on the line were my family, my home. They were closer to me than I can say, closer than any friends had been or ever could be. They had never let me down, and I couldn't do it to them. Men, I now know, do not fight for flag or country, for the Marine Corps or glory or any other abstraction. They fight for one another. Any man in combat who lacks comrades who will fight for him, or for whom he is willing to die, is not a man at all. He is truly dammed.[20]

Motivating Soldiers to Kill

Most humans have an aversion to killing other humans. In interviews of thousands of soldiers in more than 400 infantry companies that fought in the Second World War, only 15 to 20 percent of the riflemen said that they fired at the enemy, and many of those who didn't fire, nevertheless were willing to risk great danger to themselves to rescue their wounded comrades, to supply them with ammunition and to run messages, but they wouldn't fire at the enemy. According to one observer, the resistance to killing was so strong that some of the soldiers would rather die than kill an enemy.[21] Benjamin McIntyre, who witnessed a nighttime firefight at Vicksburg during the American Civil War in 1863, observed, "It seems strange that a company of men can fire volley after volley at a like number of men at not over a distance of 15 steps and not cause a single casualty. Yet such were the facts in this instance."[22] Artillery fire, which is not directed at a specific individual, accounts for more than half of the casualties on the battlefield.

Dehumanizing the Enemy. Because most humans find it difficult to kill another human being, armies often dehumanize their enemy. During war, the enemy is characterized as inferior, contemptible, as being "not like us." During the 2003–2011 war in Iraq, American soldiers referred to their prisoners as "the bad guys who want to kill you," or simply as "it." Greeks and Romans referred to everyone else as "barbarians"—foreigners or those who babbled in a different language. During World War I, the British called their enemy "Gerries"—the first syllable of the word "German"—or "Krauts," or "the Hun," and the Germans referred to their enemies as *untermench*—subhuman. During the Vietnam War, the American GIs called the Vietnamese "slopes" and "slants," and in the Korean War, North Korean soldiers were referred to as "Gooks." American soldiers are trained to refer to those they are fighting as "the enemy"—a nonspecific, collective noun, meaning "those whom we are fighting." It is easier to kill "the enemy" than to kill an Iraqi, an Afghani, a Vietnamese, or a Korean.[23] During the Mongol invasions of China in the fifth and sixth centuries CE, the Chinese characterized their enemies as "having the faces of humans, the hearts of wild beasts, creatures for whom armed robbery of the Chinese was the natural way of life."[24] During the genocide in Darfur, a member of the Janjaweed Arab militia told a black woman, "You Blacks are like monkeys. You are not human. . . . We can do anything we want to you. You cannot live here," and then he raped her.[25]

War Fever

In spite of the high costs of war, publics usually enthusiastically support their countries' wars, at least during the initial phases of a war. During

Table 6.1

Initial Public and Congressional Support for Deployment of U.S. Troops Abroad

War	Year	Public` Support	Senate Yea/Nay	House of Represenatatives
Korean	1951	77 %	n.r.*	n.r.*
Vietnam	1963	70	88/2	416/0
Persain Gulf	1991	77	52/47	250/18
Iraq	2003	76	77/23	297/133

n.r. – The Korean War was a United Nations Police Action and Congress did not vote on a war resolution

the last week of July and the first week of August in 1914, war fever spread throughout Europe as patriotic crowds in London, Berlin, Vienna, Paris, and St. Petersburg demonstrated in support of their governments' decisions to go to war and young men eagerly volunteered to enlist. In England, more than 478,000 young men enlisted in the British Army through the first week of September, and in Germany, more than 1.3 million youths volunteered for the German Army by mid-August 1014.[26] In Paris, some reservists expressed their disappointment in not being the first to be called: One said, "I'm the eleventh [day]"another replied, "You'll never make it to Berlin at that rate, it'll be over by then."[27] The French have a phrase for it, *rage militarie*, meaning the patriotic passion for war.

Table 6.1 shows the initial public support for committing the United States to war in the four wars that the country has fought since since 1945: the Korean War in 1950; the Vietnam War in 1963, the Persian Gulf War to remove Iraqi troops from Kuwait in 1991, and the Iraqi War in 2003. The table also shows the votes on the war resolutions in the US Senate and House of Representatives. In every case there was overwhelming puiblic and congressional support for going to war.

Although public support for going to war may initially be high, support for war m usually decreases during its course as the casualties

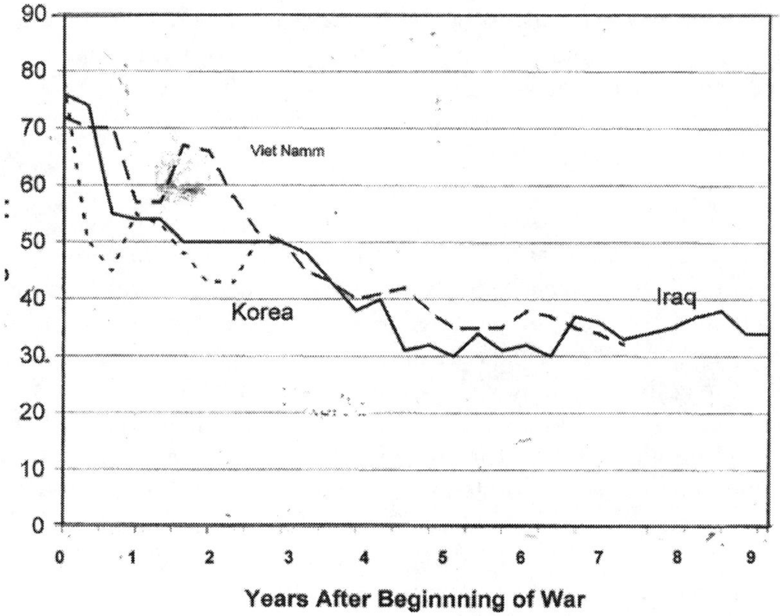

Figure 6.2 - US Public Support for the Korean, Vietnam, and Iraq Wars Is a Function of Time[28]

rise and publics become war-weary. Figure 6.2 shows how US public support has decreased during the course of the Korean, Vietnam and Iraq Wars. Much of this decline was because publics are usually overconfident of the capabilities of their militaries and underestimate the strength and determination of their enemies. In the summer of 1914, every country in Europe thought that the war would be over in weeks: "The boys will be home before Christmas." Instead, World War I lasted more than four years and left all the belligerents devastated. It took years for Europe to recover from that war. These issues of the consequences of over confidence and the willingness to take risks will be explored in Chapter 10. However the aversion to and the frustration of war rapidly disappear after a war, and patriotism and jingoism rapidly remerge. After their defeat in 1918, the German people supported Hitler's aggressive foreign policies and jingoism when he occupied the Rhineland in 1936, Czechoslovakia's Sudetenland in 1938, and attacked Poland in 1939. It was only eleven years between the end of the Korean War, which ended in a stalemate, and the Gulf of Tonkin Resolution which authorized sending combat troops to Vietnam, it was sixteen years between the US defeat in Vietnam and its invasion of Iraq in 1991, and ten years between the end of the Persian Gulf War and the invasion of Afghanistan in 2001. Thus, war seems to be a permanent institution interspersed with brief periods of peace.

Suicide Missions

The concept of self-sacrifice has long been a part of war. Many men are willing to sacrifice themselves for their country, religion or their comrades. One of the earliest recorded incidents of willful suicide was that of the Jewish Zealots at Masada in 73 CE. The Zealots were members of a Jewish political movement that sought to incite the Jews in Palestine to rebel against the Romans when they tried to introduce the imperial Roman cult that deified the emperor into Judea After the destruction of Jerusalem and the Second Temple in 70 CE, 960 Zealots took refuge at the Roman fortress at Masada, which they held against the Tenth Roman Legion for three years. Finally, in 73 CE, after the Romans were unsuccessful in their attempt to storm the Masada, they burned its walls, but when they stormed the fort to capture the Zealots, they found that the Zealots had all committed suicide by leaping off the precipice rather than become Roman slaves.

During World War II, the Japanese trained *kamikaze* (Divine Wind) pilots to fly their aircraft as manned bombs into their targets. Kamikaze strikes were used mainly against US aircraft carriers. During the war, a total of 3,843 pilots sacrificed themselves on such missions, in which they damaged or sank at least 375 US naval vessels, killed 12,300 American servicemen, and wounded another 36,400 men.[29] The Japanese Navy also used both one- and two-man pioleted torpedoes which were modified verions of conventional torpedoes, on suicide missions. In the two-man torpedoes, after aiming their *kaiten* at its target, the crew members would embrace and shoot each other in the head. Honorable suicide (*seppuku)* was embedded in Japanese culture as part of the samurai tradition, and it was indoctrinated into Japanese soldiers, sailors, and pilots. After their defeat at Saipan in the summer of 1944, hundreds of Japanese soldiers and civilians jumped from cliffs to their deaths rather than surrender to the Americans. During the Battle for Berlin in 1945, the German Luftwaffe flew "self-sacrifice missions": (*Selbstopfereinsatz*) against bridges built by the Soviets over rhe Oder River using any aircraft that was available. In April 1945, the Luftwaffe claimed that 17 bridges had been destroyed in such missions.

The first contemporary example of a suicide bombing occurred in 1972 at Tel Aviv's Lod Airport in Israel when three members of the Japanese Red Army—a terrorist organization created in 1969 and allied with the Popular Front for the Liberation of Palestine (PFLP)—used grenades and automatic rifles to kill 26 people and wounded many more. The widespread use of suicide bombers began in 1981 during Lebanon's civil war and in Sri Lanka with the Liberation Tigers of Tamil Eelam (LTTE), or the Tamil Tigers. Since then, the tactic has spread to dozens of countries; those hardest hit are Israel, Sri Lanka, Lebanon, and Iraq.

Beginning in December 1981, the Islamic Dawa Party bombed the Israeli embassy in Beirut, Lebanon, which was followed in April 1983 by Hezbollah's bombing of the US embassy in Beirut and by attacks on the US Marine and French army barracks in Lebanon in October 1983 in which 241 US Marines and 58 French paratroopers were killed. Other parties in Lebanon's civil war were quick to adopt the tactic. By 1999, Hezbollah, the Amal Movement,* the Ba'ath Party and Syria's Socialist Nationalist Party had carried out more than 50 such attacks. Although Lebanon saw the first widespread use of the suicide attacks, it was the Tamil Tigers in Sri Lanka who perfected the tactic. Their Tamil Black Tigers unit was responsible for 76 of the 168 suicide bombings between 1987 and 2005.[30] Their victims include former Indian Prime Minister Rajiv Gandhi and Sri Lanka's President Ranasinghe Premadasa. In Turkey, the Kurdistan Workers' Party (PKK) has used suicide bombers since 1984 in an effort to establish an independent Kurdish state on Turkish territory.

In 1994, two separate Palestinian groups, Hamas and the Islamic Jihad, began to conduct suicide missions against Israel. At the start of the second Intifada in 2000, they were joined by the Al-Aqsa Martyrs Brigade and the PFLP. Since the beginning of the first Intifada in 1987 until January 2007, thirteen suicide attacks have been launched against Israel, in which there were more than 900 fatalities and more than 5,000 injuries.[31]

The most dramatic suicide attacks were the September 11, 2001 attacks against the World Trade Center in New York and the Pentagon in Washington, DC, in which al Qaeda terrorists hijacked four large commercial passenger jetliners and flew two of them into the twin towers of the World Trade Center and one into the Pentagon in Washington DC, killing all those aboard the aircraft and thousands of others. The fourth aircraft crashed in Pennsylvania killing all aboard when its passengers attempted to gain control of the aircraft from the hijackers. Its target was believed to be either the White House or the Capitol building. Because the aircraft were fully fueled for their cross-country flights, they were the most powerful suicide bombs ever detonated..

The Motivations of Terrorist Campaigns. Nearly all suicide missions have been part of organized campaigns whose objectives have been secular and political rather than religious. Usually, the organizations that sponsor and organize suicide campaigns are small and weak compared to their adversaries and do not have sufficient capabilities to fight a conventional military campaign. Daud, a Hezbollah leader in southern Lebanon, explained in 1985, "We are prepared to sacrifice our lives—literally blow ourselves up in opposition to their tanks. . . . Since we cannot fight the enemy with their weapons, we have to sacrifice our lives. And this is what is happening right now in South Lebanon."[32] Sayeed Siyam, a Hamas leader in Gaza, explained, "We in Hamas consider suicide bombings inside [Israel's] 1948 borders to be the card that Palestinians can play to resist the occupation. . . . We do not own Apache helicopters ourselves, so we use our own methods. Given the methods used by the Israelis, we consider the door to hell is open. Their assassination policy and the bombardment—all this theater of war inside Palestine villages and homes—we respond to that by seeking to make Israelis feel the same, insecure inside their homes."[38] The objective of these suicide campaigns is to instill an unacceptable level of fear and terror into the enemy so the enemy will bow to the demands of the bombers' sponsors. One member of Hamas said, "We paid a high price when we used only slingshots and stones. We need to exert more pressure, make the cost of occupation much more expensive in human lives, that much more unbearable"[34].

The central objective of most of these campaigns has been to coerce a foreign state to withdraw its forces from what the terrorists. believe to be their homeland, or to establish an independent state on the territory of an existing state.[35] The bombings in Lebanon were to force the Americans, French, and Israelis to withdraw their military personnel from Lebanon. In Sri Lanka, the objective of the LTTE was to force the Sri Lankan government to allow the Tamils to form an independent Tamil state.

Prior to the 1991 Gulf War, the United States had maintained a small military mission in Saudi Arabia of between 600 and 800 personnel. In 1991, when the United States launched its offensive from Saudi Arabia to expel the Iraqi Army from Kuwait, with the conent of the Saudi government, US troop levels in the Saudi kingdom increased to almost 32,000. Since then, the United States has maintained a much larger force in Saudi Arabia than before the war. There were over 7,000 troops in the Saudi kingdom in June 1996 when al Qaeda attacked the Khobar Towers, a housing complex for US military personnel in Saudi Arabia. Osama bin Laden justified the attack in his infamous fatwa against the United States, "The ruling to kill Americans and their allies—civilians and military—is an individual duty for every Muslim who can do it in any country in which it is possible to do it, in order to liberate the al-Aqsa Mosque [in Jerusalem] and the holy mosque [in Mecca] from their grip, and in order for their armies to move out of all the lands of Islam, defeated and unable to threaten any Muslim."[36]

The terrorists have reason to believe that their campaigns to convince a nation to withdraw its forces from a Muslim country can be successful. On March 10, 2004, when Spain had 1,300 troops in Iraq, ten terrorist bombs killed 191 commuters and wounded more than 1,800 others as four commuter trains were departing from three stations in Madrid. Four days after the attacks, Spain elected a socialist prime minister, José Luis Rodriguez Zapatero, who had promised to withdraw Spain's troops from Iraq, and after his election, he promptly did what he had promised. Three weeks after the attacks, the Spanish police surrounded the apartment building in a Madrid suburb where the terrorists were hiding, and as the police entered the apartment, the terrorists detonated a bomb that killed the five terrorists and one of the police officers.[37]

Three other European countries then followed Spain's lead. The Netherlands withdrew its 1,400 troops in March 2005; Italy withdrew its 3,000 troops in September 2005, and Bulgaria withdrew its 450 troops in January 2006. Although Great Britain had been reducing its troops in Iraq from a peak of 46,000 after the invasion to about 7,000 in July 2005, terrorists detonated bombs on three subway trains in London and on a bus on July 7, 2005. Britain then withdrew all but 2,500 of its troops from Iraq.

In Israel, the terrorists' objective is to force Israel to vacate the West Bank that has been occupied by Israel since 1967, and to force Israel to allow the establishment of an independent Palestinian state The Palestinian Hamas movement that rules the Gaza Strip has taken a rejectionist position that denies Israel's right to exist. Hamas and the Islamic Jihad believe that Allah endowed all of Palestine to the Muslims, including the territory of the state of Israel, just as Jews believe that God endowed Palestine to them. Muslims claim that the Zionists are foreign invaders supported by the Crusaders (Christians) in Europe and America. They also believe that Allah's sacred endowment cannot be negotiated away.

Since the Palestinians are incapable of inflicting a conventional military defeat on Israel, their strategy has been to inflict as much pain as possible on Israel by killing as many Israeli civilians and children as possible. The Palestinians claim that they are engaged in a holy war against an occupying power and that their religion and international law permit the use of any means of resistance to the occupation; they are retaliating against Israel killing members of armed groups and Palestinians generally; that striking civilians is the only way they can make an impact upon a powerful adversary; that Israelis generally and settlers in particular are not civilians.[38]

Motivating Suicide Bombers What motivates a person to be willing sacrifice his life to kill others? People commit suicide for many reasons. The nineteenth century European sociologist Emil Durkheim defined three kinds of suicide: *egoistic suicide* is committed by people who are unable to cope with an intense psychological trauma and choose death as the means to escape from their painful existence; *fatalistic suicide* occurs when a person is subject to unbearable conditions such as extreme discipline, oppressive regulation, or seclusion; and *altruistic suicide* is suicide in which a person sacrifices her or his life for the benefit of others.[39] Examples of the latter category include a parent sacrificing her or his life to save her or her or his child, and a soldier throwing himself on a hand grenade to save the lives of his comrades. Most suicide terrorist acts can be classified as altruistic. Although the motives of most suicide attacks are not religious, most Muslims who carry out these attacks consider themselves to be *shaheed* (martyrs), the crucial element of which is purity of motive, or doing Allah's will, rather than acting in one's own self interest. Most of the *shaheed* volunteer for their missions and are not coerced to do so. One Palestinian volunteer said, "In the spring of 1993, I began to pester our military leaders to let me do an operation. . . . I wanted to do an operation that would incite others to do the same."[40] Sheikh Ahmed Yassin, the spiritual leader of Hamas, explained, "Love of Martyrdom is something deep inside the heart. But these rewards are not in themselves the goal of the martyr. The only aim is to win Allah's satisfaction. This can be done in the simplest and speediest manner by dying in the cause of Allah. And it is Allah who selects the martyrs." There is no shortage of willing recruits. A Hamas leader explained, "Our biggest problem is the hordes of young men who beat on our doors, clamoring to be sent. It is difficult to select only a few. Those whom we turn away return again and again, pestering us, pleading to be accepted."[41]

In preparing a person to be a *shaheed*, a member of Hamas said, "We focus his attention on Paradise, on being in the presence of Allah, on meeting the Prophet Mohammad, on interceding for his loved ones so that they, too, can be saved from the agonies of Hell, on the heavenly virgins, and on fighting the Israeli occupation and removing it from the Islamic trust that is Palestine." One volunteer who was ready to perform his mission but who was arrested before he could carry it out talked about the sense of the immediacy of paradise: "It is very, very near—right in front of our eyes. It lies beneath the thumb. On the other side of the detonator."[42]

In selecting candidates for *shaheed*, one planner for the Islamic Jihad said that his organization carefully scrutinizes their motives: "We ask this young man, and we ask ourselves, why he wishes so badly to become a human bomb. What are his real motives? Our questions are aimed at clarifying first and foremost for the boy himself his real reasons and the strength of his commitment. Even if he is a long time member of our organization and has always wanted to become a *shaheed*, he needs to be very clear that in such an operation there is no drawing back. Preparation bolsters his conviction, which supports his certitude. It removes fear."[43] During the week before the operation, two "assistants" are delegated to stay with the candidate martyr at all times. They report any signs of doubt, and if the young man seems to waiver, a senior trainer will arrive to bolster his resolve. The father of one *shaheed*, who, with his friend, carried out an explosion in Beit Lid in Israel in 1995, said with pride, "Even after the friend saw his son ripped to shreds, he did not flinch. He waited before exploding himself in order to cause additional deaths."[44]

In preparation for their mission candidates undergo an intensive spiritual preparation. The candidate is instructed to read six specific chapters from the Quran that address such themes as jihad, the birth of the nation of Islam, war, Allah's favors, and the importance of faith. He undergoes religious lectures for two or three hours each day, undertakes lengthy fasts, and prays for most of the night before the mission. He must pay all of his debts and ask forgiveness for his offenses. Finally before he undertakes his mission, the *shaheed* prepares his will on paper, an audiocassette, or a videotape, sometimes on all three. The videos are shot against the background of the sponsoring organization's banner and slogans and show the *shaheed* reciting from the Quran, posing with guns and bombs, exhorting his friends to follow his example, and extolling the virtues of jihad: "This is my free decision, and I urge you to follow me." The *shaheed* watches his video and those of his predecessors, which prepares him to confront death, not to fear it. Just before his mission, he performs a ritual cleansing of his body, puts on clean cloths, recites the traditional Muslim battle prayer. He puts a Quran in his left breast pocket and straps the explosive belt around his waist.[45]

The Palestinian *shaheed* does not require any additional training or preparation for his mission beyond that described above. His weapons, suicide belts, vests, and satchels are supplied by the sponsoring organization. In Western Europe and North America, however, it is difficult for a would be *shaheed* to import his weapons, so he must make them himself, so the terrorists must be trained to make their own weapons. Al Qaeda and other terrorist organizations have established training facilities in Pakistan, Afghanistan, Iran, Iraq, and Palestine, and there have been reports of a facility in Italy. The leader of the July 7, 2005, London attacks, Mohammad Sidique Kahn, learned how to make improvised explosive devices while attending military training camps in Pakistan and Afghanistan. He is also believed to have participated in the planning of an attack on Mike's Place, a bar in Tel Aviv. A second participant in the bombings, Shehzad Tanweer, had traveled to Pakistan with Kahn in 2004, where he studied at a *madrasa*—a Muslim religious school. Prior to the bombing, he made a video in which he stated, "What you have witnessed now is only the beginning and will become stronger until you pull your forces out of Afghanistan and Iraq and until you stop your financial support to America and Israel."

Chapter 7

The Human Superorganism

> Thus was born the human condition, selfish at one time,
> selfless at another, the two impulses often conflict.
>
> Edward O. Wilson[1]

> Thre are no Robinson Crusoes
>
> Charles Malik[2]

In the previous chapters, we have seen that war is an integral part of man's social life and that it coevolved with his culture. In this chapter, we will begin our exploration of how this condition could have evolved; and we will see that human societies can be treated as superorganisms that have many of the properties of and that behave as unitary biological organisms.

Superorganisms

We humans live in the largest social aggregations of any animal species. The populations of the largest nations include over a billion people. The next largest communities are ant colonies which seldom exceed a few million ants,[3] and the colonies of all other social insects are much smaller, usually less than a few hundred individuals.[4] The largest group of primates that has ever been observed is a sleeping party of 750 baboons, although their troops seldom exceed a a hundred individuals.[5,6] Among the great apes, chimpanzee communities usually include between 15 aand 80 individuals; bonobo communities between 40 ands 120 individuals, and gorilla communities include between two and thirty individuals, so humans are unique among animal species in the size of their social groups.[6]

Large animal social groups, such as ant and bee colonies and some human organizations, appear to have some of the properties of a single organism and have been designated as superorganisms. In his book *The Social Insects: Their Origin and Evolution,* William Norton Wheeler designated large colonies of insect species as superorganisms: "We have seen that the insect colony or society may be regarded as a superorganism, and hence as a living whole bent on preserving its moving equilibrium and integrity."[7] Previously, Wheeler had written in his essay "The Ant Colony as an Organism," "The most general organizational character of an ant colony is its individuality. Like the cell or the person, it behaves as a unitary whole, maintaining its identity in space, resisting its dissolution and, as a general rule, any fusion with other colonies of the same or alien species."[8]

At that time in the early twentieth century, only haplodiploid species[*] of insects such as ants, bees, wasps, and thrips were considered to form superorganisms because of their unique reproductive system, but the concept of a superorganism was later expanded to include beetles and still later to include two mammalian species: the naked mole rat and the Damaraland mole rat. The common features of all these species are that they are eusocial (truly social, *eu* = true, Greek) because they live in colonies in which: (1) the adults are divided into reproductive and nonreproductive castes,[†] (2) there are overlapping generations of adults, and (3) there is a division of labor among the adults.[9] Some authors such as Bert Hölldobler and Edward O. Wilson[10] maintain that eusociality is a necessary and sufficient criterion of a superorganism, whereas others, including myself, believe that while eusocial species form superorganisms, the category of superorganism should be expanded to include a broader category of species. Here, I propose the following definition of a superorganism:

- Its members are unable to exist independently for extended periods;

---

[*] In haplodiploid species one sex (usually the males) have only one set of chromosomes, whereas the other sex (usually the female) has two sets of chromosomes. In diplodiploid species, both sexes have two sets of chromosomes

[†] In these species the queen emits phemerons (odors) that prevent the other females of the colony from reproducing

- T he superorganism can be self-sufficient;
- T here is a division of labor among its members;
- T here is a high degree of cooperation among its members;
- Intelligence is distributed among its members;
- It is willing to sacrifice its members for its survival and aggrandizement
- Its members are willing to sacrifice themselves for the survival and aggrandizement of the group.

Under this definition both eusocial species and some huuman societies qualify as superorganisms.

The Human Superorganism

That a human society can be a superorganism was first proposed by the 19th-century scientist/philosopher Herbert Spencer, who entitled the first chapter of his *Principles of Sociology* "Superorganic Evolution," in which he noted that as human societies evolved, they increased in both size and their development, and that the divisions of society are analogous to the complexity in the same way that living organisms do. during the course of their development, and that the divisions of society are analogous to the organs of a living body:

> It is also a character of social bodies, as of living bodies, that while they increase in size they increase in structure. A low animal, or the embryo of a high one has few distinguishable parts; but along with its acquirement of greater mass, its parts multiply and simultaneously differentiate. It is thus with a society. At first the unlikenesses among its groups of units are inconspicuous in number and degree, but as it becomes more populous, divisions and subdivisions become more numerous and more decided. Further, in the social organism, as in the individual organism, differentiations ceases only with that completion of the type which marks maturity and precedes decay.[11]

The only human institution that can qualify as a superorganism under the above definition is a state, defined as a group of humans who occupy and controls a defined territory, and has a centralized government that maintains a monopoly on the legitimate use of force to enforce its authority and to defend and extend its territory and influence. Until the 20th century, some chiefdoms—such as North American Indian tribes, the Polynesian island communities, the Zulu tribes in South Africa, and the Nuer and Dinka tribes of South Sudan, which were autonomous political units comprising a number of tribes, villages, or communities under the permanent control of a prominent chief—could be considered to be superorganisms.[12] By the early 20th century, however, nearly all independent chiefdoms had become subservient to larger states.[13] Unless it constitutes a state or is totally self-sufficient, neither an ethnic nor religious group can be considered to be a superorganism.

Very few humans live as solitary hermits, and almost all are highly dependent on the state in which they live for their existence. Although most states can be self-sufficient, they frequently cooperate with one another through trade and military alliances, and in other areas of mutual interest. In all states, there is a division of labor among members into highly specialized occupations, but there is also flexibility in their employment. A state can conscript its citizens to fulfill essential duties, such as military service and jury duty, and can punish its citizens who violate its laws and norms. No single individual holds all of the knowledge that is needed for a functioning human superorganism; hence knowledge is distributed among the members of the state. Humans are highly cooperative and enthusiastically work together to achieve common goals. Almost all states have fought wars in which they willingly sacrifice their citizens for the state's survival and aggrandizement, and many of their citizens volunteer to serve in their armed forces.

Most humans have values and priorities that supersede their own survival: that the survival and aggrandizement of their state, ethnic group, nation, country, or religion is more important to them and takes precedence over their own survival and well-being. This implies that the fundamental unit of human existence is not necessarily the individual but the social unit to which she or he belongs. At this level, individuals are willing to take existential risks, and the superorganism or state is willing to sacrifice its members to its own survival and aggrandizement. Walter Lippmann once observed that men don't become happy by satisfying their desires but by living within the belief system of their culture, which restrains and gives coherence to their lives: "Above all the other necessities of human nature, above the satisfaction of any other need, above hunger, love, pleasure, fame—even life itself—what a man needs most is the conviction that he is containedwithin the discipline of an ordered life."[14]

Unlike insects, however, humans as individuals have needs and desires that are independent of their society, and in this sense, they are selfish or self-oriented instead of being group-oriented. Thus human societies are faced with the problem of balancing the desires of the individuals with those of the society as a whole—the problem of the individual's will versus the collective will. Herbert Spencer recognized this when he wrote,

> [T]he lives of all parts of an animal should be merged in the life of the whole, because the whole has a corporate consciousness capable of happiness or misery. But it is not so with a society; since its living units do not and cannot lose individual consciousness. This is an everlasting reason why the welfare of citizens cannot rightly be sacrificed to some supposed benefit of the State, and why, on the other hand, the State is to be maintained solely for the benefit of citizens. The corporate life must here be subservient to the lives of the parts, instead of the lives of the parts be subservient to the corporate life.[15]

Spencer, however, later qualified this view when he discussed the question of discipline during warfare in the third edition of his *Principles of Sociology*,

> Habitual war, requiring prompt combination in the action of parts, necessitates subordination. Societies in which there is little subordination disappear, and leave outstanding those in which subordination is great; and so there are produced, societies in which the habit forced by war and surviving in peace, brings about permanent subordination to a government."[16]

Thus human societies are faced with the problem of balancing the desires of their members with the needs of the society. The degree of individual freedom versus the control that the state maintains over its citizens varies greatly among states. Totalitarian states such as dictatorships, communist states, and some theocracies severely restrict the freedoms of their citizens compared with western liberal democracies. Among such democracies, however, there is a wide range of opinion regarding the extent of individual freedoms: Libertarians believe that the authority of the state should be strictly limited and that individual freedoms should be maximized, whereas social democrats believe that the collective happiness and well-being of individuals is maximized when the state redistributes the wealth of its citizens to obtain a maximum of social and material benefits for all of its citizens.

Like all biological organisms, human superorganisms are born, mature, and then die and become extinct. History contains a graveyard of civilizations, states and empires that have lived magnificent lives and then have died and disappeared from history. How did such superorganisms emerge and evolve?

The Evolution of the Human Superorganism[17]

The first step in the evolution of the human superorganism was the formation of groups, which is likely to have begun in the era of *Pan prior*, the common ancestor of humans and chimpanzees that lived between approximately between 8 million and 6.5 million years ago, as described in Chapter 2. At first, the members of these groups were loosely bonded together, similar to the way contemporary chimpanzee communities are, but group cohesion slowly tightened over time. The first *P. prior* groups probably formed because of the advantages that cooperative living afforded them in foraging for food, and the protection of females and their offspring from sexual and other predators in exchange for providing dependable sexual partners for the males. Group living also increased the fecundity of the females.

The next step in the evolution of the human superorganism was the assimilation of these emerging traits into the hominin genome. The males foraged and hunted together, shared food with and assisted the females in caring for their offspring, They took existential risks in hunting and in defending their group. Other traits, such as the acquisition of territory, the construction of defensible home sites, the development of a common set of cultural norms, and possibly the emergence of a set of religious beliefs fostered the emergence of a superorganism. Over time, parochial altruism developed and became fixed in the genomes of the members of these groups as described in the previous chapter.

Such primitive hunter-gatherer groups, however, could not be considered to be superorganisms because the human superorganism did not emerge until the formation of large societies during the agricultural revolution, when agricultural surpluses make it possible for some individuals to engage in occupations other than farming such as craftsmanship, building, trading, the priesthood, soldiering, and bureaucracy. As agricultural communities grew, humans had to subordinate themselves to the impersonal authority of their rulers and to interact with more people on an impersonal basis. Thus fully developed states and civilizations emerged as human superorganisms.

Group Selection and the Emergence of War

Although others before him recognized that man is a social animal, Charles Darwin was the first to elaborate on the nature of man's sociality from a evolutionary viewpoint. He recognized that for primitive humans who lived in tribes, their loyalties did not usually extend beyond their tribe—that they were parochial altruists—but he also recognized that humans are torn between satisfying their own selfish desires and the need to be part of and to promote the interests of their tribe. Darwin thought that man's social instincts might not be stronger than his "instincts for self preservation, hunger, lust, vengeance &c."[20] Darwin also had a conception of altruism*[18] when he wrote,

> The bravest of men . . . who freely risk their lives for others, would on the average perish in larger numbers than other men. Therefore, it seems scarcely possible . . . that the number of men gifted in such virtues, or that the standard of their excellence, could be increased through natural selection, that is, by the survival of the fittest.[19]

To solve this conundrum, Darwin conceived that natural selection could also work on groups of individuals as it does on individuals within a group. Just as natural selection can select between individuals in an isolated population on the basis of traits that cause them to be more fit in their environment, and displace those who are less fit, and thus cause the population to change so that it is more adapted to its environment as described in the Appendix, natural selection can also act on a population of groups to select those groups that possess the most beneficial traits for their survival.

* The word altruism was first used by the French philosopher-psychologist Agueste Comte (1798–1857) in his 1853 work, *The Philosophy of Augueste Comte*, but evidently, Darwin was unaware of Comte's work.

Lethal conflict between groups is one of the mechanisms that selects between groups. Those groups whose members possess genes that cause them to be the strongest and the most able to defeat other groups that they fight will prevail. Charles Darwin recognized this when he wrote in *The Descent of Man and Selection in Relation to Sex,*

> When two tribes of primeval man, living in the same country, came into competition, if the one tribe included (other circumstances being equal) a greater number of courageous, sympathetic, and faithful members, who were always ready to warn each other of danger, to aid and defend each other, this tribe would without a doubt succeed best and conquer the other. Let it be borne in mind how all-important, in the never-ceasing wars of savages, fidelity and courage must be. The advantages which disciplined soldiers have over undisciplined hordes follows chiefly from the confidence which each man feels in his comrades. . . . Selfish and contentious people will not cohere and without coherence nothing can be effected. A tribe possessing the qualities in a high degree would spread and be victorious over other tribes; but in the course of time it would, judging from all past history, be in turn overcome by some other and still more highly endowed tribe. Thus the social and moral qualities would tend slowly to advance and be diffused throughout the world.
>
> There can be no doubt that a tribe including many members who, from possessing in a high degree the sprit of patriotism, fidelity, obedience, courage and sympathy, were always ready to give aid to each other and to sacrifice themselves for the common good, would be victorious over most other tribes; *and this would be natural selection.*[20] (italics added)

Thus, Darwin formulated a theory of group selection: that nature selects between groups on the basis of their ability to survive over other groups, a concept that would be rejected and then revived in the late 20th century. Darwin, however, saw a flaw in his argument when he tried to reconcile it with his theory of natural selection, "But it may be asked, within the limits of the same tribe [how] did a large number of first become endowed with these social and moral qualities, and how was the standard of excellence raised?"[21] In other words, how did the cooperation among the members of a tribe evolve within a group of otherwise selfish individuals? The questions of how selection can lead to cooperative behavior and group selection have occupied evolutionists for many years. Suppose that there are two alleles* of a gene; one for cooperative behavior, and one for selfish behavior, then the condition for the cooperative allele to increase in a population and to eventually replace the selfish allele is for the cooperative allele to be the dominate one for the individuals who carry both alleles and for individuals who carry the cooperative allele to reproduce at a higher rate than those who carry the selfish allele.

In spite of Darwin's convincing arguments of the existence of group selection discussed above, the issue of how group selection works has been debated among evolutionists since the 1960s and many mathematical models of group selection have been proposed,[22,23,24,25,26,27,28] no single explanation of group selection has been universally accepted, although the existence of group selection has been demonstrated experimentally.

The Human Superorganism and Western Tradition

The ideas expressed above—that humans are more committed to the survival and aggrandizement of their country, religion, or ethnic group more than they are to their own survival and selfish desires—may be disturbing to some readers. Bear in mind, however, that before the European Enlightenment of the 17th and 18th centuries, nearly all societies were ruled by autocrats and ordinary men were obliged to obey their feudal lord, king, or emperor, to serve in his army, to pay his taxes, and in many cases, to provide corvée labor to his lord. Many of these autocratic rulers deified themselves to increase their power and authority. The Egyptian pharaohs and Roman emperors deified themselves and demanded that they be worshiped by the proletariat. During the Reformation, Protestant rulers in Germany and Bavaria replaced the images of the saints in their churches with their own and demanded that their subjects to worship them instead of the saints. In England, Elizabeth I made attendance at Anglican religious services mandatory for her subjects and demanded that they worship

---

* An allele is a variant of a gene. The gene for eye color has alleles for brown, blue and gray eyes.

her.. After Ferdinand and Isabella expelled the Jews and Muslims from Spain in 1492 and 1502 respectively, they instituted the Spanish inquisition to consolidate their power in Spain and to confer legitimacy on their new Spanish state.

This relationship between the individual and the state began to change in the the middle of the 17th century. Between 1517 and 1618 CE, Europe had been engaged in the Protestant Reformation that had spawned the religious wars which engulfed all of Christian Europe and culminated in the Thirty Years War (1618–1648 CE)—the most devastating European war to that time. Germany and Austria lost more than a third of their populations, which fell from 21 million to 13 million; Bohemia lost 80 percent of its population, which fell from 3 million to 800,000, and of the 35,000 villages that had existed in Bohemia in 1618, 29,000 were deserted in 1648.

The Enlightenment or the Age of Reason that followed was a reaction to the carnage of the war, to absolute monarchy, and to religious persecution. It promoted the desire of men to rule themselves, and the separation of church and state. It was called the Age of Reason because its proponents in France, the Netherlands, Germany, and England and its American colonies believed that the universe was governed by predictable laws of nature, and because the scientific world view that they had adopted embodied a rational perspective free of religion and superstition. For Emmanuel Kant, the Enlightenment was "man's final coming of age, the emancipation of human consciousness from the dogmatic state of ignorance and error."[29] For the Dutch philosopher Baruch Spinoza, the Enlightenment was freedom of thought, expression, and the press; individual liberty; the elimination of religion from the legislative process and education; and the separation of church and state.

In 1690, the English philosopher John Locke wrote, "Hence it is evident that absolute monarchy, which by some men counted as the only government in the world is inconsistent with civil society, and so can be no form of civil government at all."[30] The Enlightenment put the individual above the state and made the purpose of government to serve the individual. These ideas were the basis of the democratic revolutions in Europe and the western hemisphere that began with England's Glorious Revolution in 1688, and included the American and French Revolutions, and then spread through South America in the 19th century, and then through East Asia after World War II

The ideas of the Enlightenment are expressed most succinctly inthe American Declaration of Independence: "We hold these truths to be self-evident: That all men are created equal, that they are endowed by their creator with certain unalienable Rights, among which are life, liberty and the pursuit of happiness," and in the 1688 English Bill of Rights, the French Rights of Man, and the American Bill of Rights.

Where the Enlightenment places the state subservient to the individual, the human superorganism places the individual subservient to his state, religion, and ethnic group, thus creating a tension between the state and the individual. Whereas the superorganism is the result of biological natural selection, the Enlightenment was a cultural revolution, and it is not surprising that the two are in conflict. Man's noblest aspirations are more often than not thwarted by his base biological instincts. As Edward O. Wilson has remarked, "An unavoidable and perpetual war exists between honor, virtue and duty, the products of group selection, and selfishness, cowardness and hypocrisy, products of individual selection."[31]

Democratic governance, however, is unique to western civilization and has not been easily adapted to other civilizations. It grew out of a particular milieu of conflict and authoritarianism that existed in Europe prior to 1650CE. Although many non-western nations have taken on the trappings of democracy, there are only a few truly democratic countries outside of Europe and the western hemisphere; they include Japan, South Korea, Taiwan, Australia, South Africa, and India. Although several Muslim countries have tried to install democratic governments, none have truly succeeded. Most non-western states have governments of a single authoritarian party, and there is no conception of the separation of religion and the state. Some countries, such as Russia and China, view democracy as chaotic and dysfunctional, and a majority of their citizens feel most comfortable with strong authoritarian governments. Turkey has been torn between its traditional Islamic culture and the democratic institutions introduced by Mustafa Ataturk in the 1920s, which were to be upheld by the Turkish army; but today, Turkey appears to have turned its back on the west and to be reverting to its earlier Islamic traditions.

Chapter 8

Aggression: A Necessary Condition for War

> Aggression evolves . . . as a contingency plan. It is a set of complex responses of the animal's endocrine and nervous system, programmed to be summoned up in the time of stress. Aggression is genetic in the sense . . . that its components have proved to have a high degree of heritability.
>
> Edward O. Wilson[1]

Aggression is one of the most universal traits among animals, including humans, and can be defined as hostile actions of an individual or a group toward an another individual or group, with the intent of causing the severe injury or death of the opponent, of driving the opponent from one's territory, or causing the opponent to cease its aggressive actions. Aggression can be directed against conspecifics or against members of other species. From this definition of aggression, war is aggression involving two or more human social groups. In Chapter 3, we saw that warfare among primitive human societies seemed to be an end in itself and, in many cultures, an essential part of the life of a male. Because aggression is common in almost all animal species, including crustaceans, fish, insects, birds, and mammals,[2] aggression must have a genetic basis that evolved in the very distant past and probably because of the need of animals to acquire high protein, energy rich animal food; to acquire and defend territory in which the necessary resources for existence are located; to acquire and defend sexual partners; to defend against both nonspecific and conspecific predators; to protect offspring against predators; and to establish dominance relationships and hierarchies in their social groups.

Aggressive behavior can be classified as either premeditated or impulsive. Premeditated aggression is planned behavior with a specific goal. Aggressive war is a form of premeditated aggression that is socially sanctioned. In contrast, impulsive aggression is characterized by high levels of autonomic or involuntary arousal that is precipitated by negative emotions such as anger or fear.[3] Impulsive aggression is usually attributed to individuals, but whole societies can engage in impulsive aggression, as the European nations did at the beginning of World War I in August 1914, as described in Chapter 6.

There are significant differences between aggression by males and females. Male aggression is usually directed toward the acquisition or protection of resources, mates, and territory, or toward gaining a dominant position in their societies, whereas female aggression is usually directed toward the protection of offsprings and nests. Lactating female mammals tend to be aggressive in protecting offspring, whereas non-lactating females are seldom aggressive.[4] Although in most species, males are more aggressive than females because of differences in their hormonal and neurological processes, in some species, such as the lion (*Panthera leo*) and spotted hyena (*Crocuta crocuta*), the male and female roles are interchanged and the females are more aggressive than the males. Female lions do most of the hunting, and female spotted hyenas are aggressive toward other conspecifics.

The Genetic Basis of Aggression

Animal breeders and farmers have known for thousands of years that both the physical and behavioral traits of animals are heritable and can be selectively bred for. The first chapter of Charles Darwin's *The Origin of Species* discusses artificial selection, which is the breeding of animals for specific traits. Dogs and other animals have been bred as much for their behavior as for their appearance. Spaniels, for example, were originally bred to creep up on birds and then to spring and frighten them into the hunter's net, but after hunters began using shotguns, spaniels were taught to point rather than to spring, and these behavioral traits became heritable in dogs. Breeding or artificial selection can be very finely tuned. In England, where dogs are primarily used for hunting, there are 26 recognized breeds of hunting dogs, whereas in France, where dogs are primarily used for farm work, there are 127 different breeds of shepherd and stock dogs.[5]

Aggression is a heritable trait and therefore has a genetic basis. The

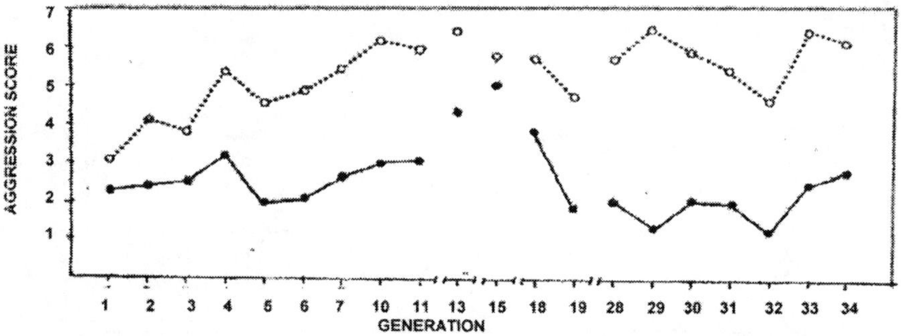

Figure 8.1. The Heritability of Aggression in Mice.[6]

heritability of aggression in animals, particularly in mice and rats, hasbeen extensively studied. More than forty years ago, Kristi Lagerspetz at the University of Turku in Finland showed that mice can be selectively bred to be either aggressive or nonaggressive.[7,8,9,10,11] In his experiments, 24 male albino mice that were not siblings were tested for their aggressiveness and then divided into two groups of the most aggressive and the least aggressive mice. Members of both groups were then mated and their offsprings were again tested for their aggressiveness and then divided into the most and least aggressive groups. The experiment was conducted for 34 generations and the results are shown in Figure 8.1.

The degree of aggressiveness of the mice was determined by placing a pair of mice from the same generation in an observation cage, observing their behavior for seven minutes and then scoring the aggressiveness of each mouse on a scale zero—the subject showed no interest in his opponent—to six—fierce wrestling and biting hard enough to draw blood.[12] The animals were tested at the age of 4½ months on seven successive days and their daily scores were averaged. During generations 12 through 15, the nonaggressive strain acquired an infectious disease, and the testing was discontinued during those generations in both strains to keep the nonaggressive strain alive. However the general trends in both strains continued after the 28th generation, clearly demonstrating the heritability of aggressiveness in mice.

Several studies of identical (monozygotic) human twins who have the identical genomes at conception, have shown that aggression in humans is heritable and thus has a genetic basis. One study by Ghodsain-Carpey and Baker showed that 94 percent of the variance in .the

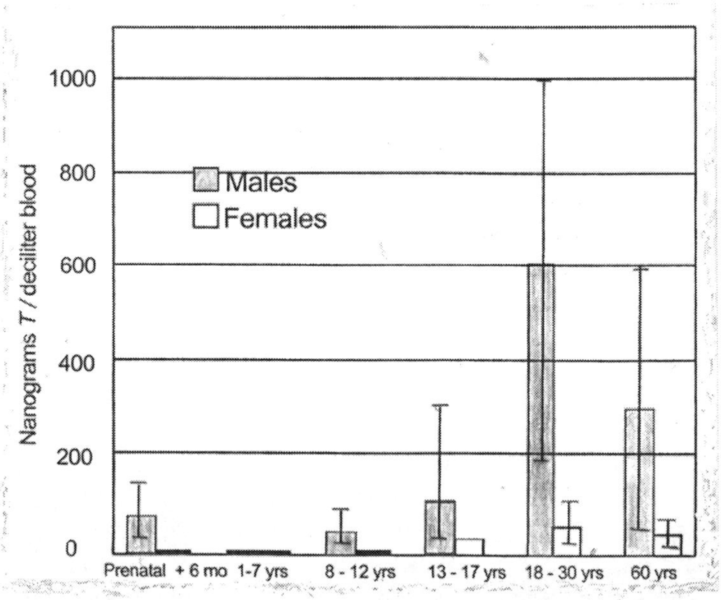

Figure 8.2. Blood Concentrations of Testosterone in Human Males and Females.[13] (Error bars show 95% interval relative to mean values)

aggressive behavior in four to seven year old identical twins can be attributed to genetic factors.[14] Two other twin studies also showed substantial genetic contribution to aggressive behavior in boys from the ages from 7 to 16 years.[15,16] Another study compared the genetic contributions to bullying in identical and fraternal (dizygotic) twins and found that the genetic contribution in the fraternal twins was roughly half of what it was in identical twins: 0.42 versus 0.72.[17] Identical twins result from the division of a fertilized egg immediately after conception and therefore have identical genomes at conception. Fraternal twins result from the independent fertilization two eggs and have on the average the same number of identical genes that normal siblings have.

Testosterone and Aggression

Aggressive behavior is one of the most consistent behavioral differences between males and females throughout the animal kingdom. In humans ten times as many men are convicted of violent crimes as are women; in almost all armies, military combat units are composed entirely of males; and at every age during their school years, many more boys initiate physical assaults than girls do. Furthermore

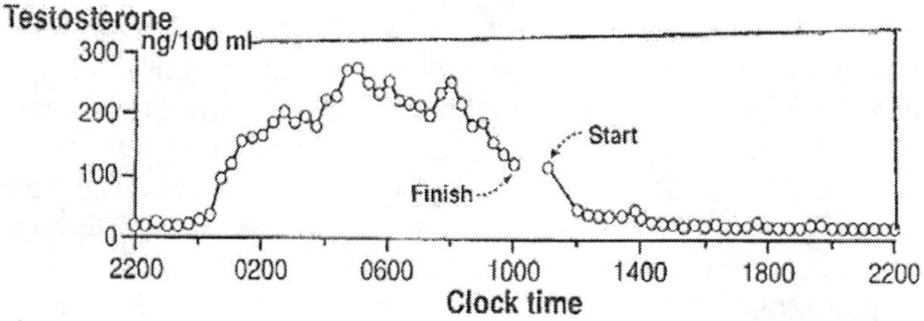

Figure 8.3. Daily Cycle of Testosterone in a 14-Year-Old Boy.[17]

boys and girls tend to express their aggression in different ways: Boys are physically and verbally more aggressive than girls and express their aggressiveness in their play fantasies and in their sports, whereas girls tend to be more aggressive socially, e.g., by excluding other girls from a group. This difference begins to develop between the ages of 2 and 2½ years and does not begin to decline until early adulthood.

The major source of these differences in the aggressive behavior between males and females are the steroid sexual hormones—the female estrogens and the male androgens. Testosterone is the most important of the male androgen hormones. During most of their lives, males have 10–20 times more testosterone in their blood as females do, as shown in Figure 8.2, Females have 3–20 times more of the estrogen hormones in their blood than males do, depending upon where a female is in her menstrual cycle. However the female ovaries produce testosterone which is immediately concerted into female estrogens.

In males, blood testosterone levels follow a circadian rhythm and can fluctuate widely during the course of a day, as shown in Figure 8.3 for a 14 year old boy. Note that the level is highest during his sleeping hours. Testosterone levels also vary between populations. Blood testosterone levels for the !Kung bushmen in southern Africa are at the low end of the range for European and North American men.[18]

The Physiology of Testosterone. The sexual steroid hormones—male androgens and female estrogens—are synthesized in the male and female sexual organs. Both males and females inherit an X chromosome from their mothers, whereas males inherit a Y chromosome from their fathers and females inherit a second X chromosome from their fathers. Slightly more than half of the human male sperm cells contain an X chromosome, and slightly less than half contain a Y chromosome, which accounts for why more females are born than males. Thus, females have two X chromosomes, one of which in each cell becomes almost completely inactive during embryonic development, whereas males possess both an X and a Y chromosome. The male Y chromosome contains a sex-determining region (SRY) gene that is the genetic difference between males and females. The SRY gene codes for the testis determination factor (TDF), a protein that causes the asexual gonads of the male fetus to develop into testes, whereas in females, who do not inherit the SRY gene, fetal gonads develop into ovaries.[19]

In male fetuses, prenatal blood testosterone levels are high during the last trimester and remain high for about the first six months postpartum and then begin to fall. During this period, testosterone is carried by the blood from the testes to the brain, where it masculinizes the male brain, stimulating the growth and the development of neurons in the hypothalamus and the amygdalae to give the neural circuits in these regions a uniquely male character, although this process is not yet completely understood.[20] A recent study has found that the genes on the Y chromosome are responsible for the phenotypic differences in the brain neurons of males and females, demonstrating that the Y chromosome is the source of differences in the male and female brains. After the age of about one year, male blood testosterone levels remain very low until the age of about eight, when testosterone levels begin to rise in males to complete the masculinization of their bodies. Testosterone levels remain high through early adulthood until the age of about 30, when they begin to decline into old age.

Testosterone and Aggressive Behavior. The relationship between blood testosterone levels and the incidence of aggressive behavior appear to be independent of the blood testosterone level, whereas the intensity of aggressive behavior appears to be correlated with blood testosterone levels.[21] High blood concentrations of testosterone found in male prison inmates who had committed violent crimes and are considered "tougher" than their peers, and in inmates who were chronically aggressive toward other inmates. The blood testosterone levels of both male and female delinquents who had histories of violent or rebellious behavior were found to be substantially higher than those of a control group of college students. Rapists who had been institutionalized for committing violent rapes also had high levels of testosterone in their blood.[22] It has been shown that the administration of the gonadotropin releasing hormone which reduces testosterone levels in men, also reduces their aggressive

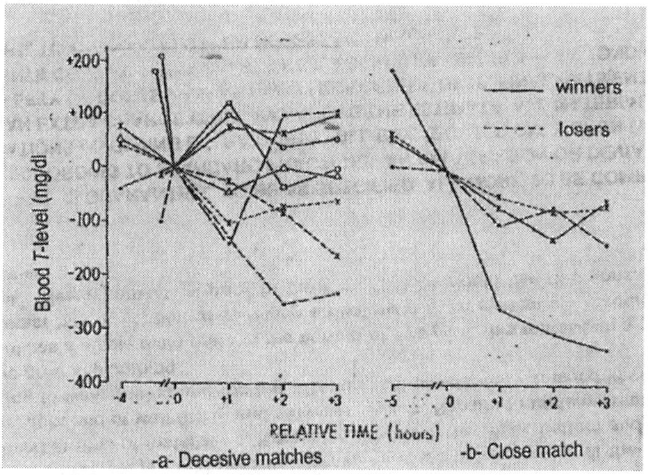

Figure 8.4. Blood testosterone Levels of Winners (solid) and Losers (broken) in a Tennis Match[23]

behavior. Athletes who use anabolic steroids have been observed to be more aggressive than those who do not use the steroids.[24,25,26]

Testosterone and Dominance. Competitive sports are a form of ritualized aggression. In most competitive sports, as in warfare, winners assume a dominant relationship with respect to losers. The relationship between aggression and testosterone appears to be an indirect one. In males, blood testosterone levels usually rise before competitive activities as if to prepare them for the competition and risk taking. After the competition, blood testosterone levels usually rise again in the winners but fall in the losers, a phenomenon called the reciprocal challenge effect or the challenge hypothesis. Testosterone levels seem to be the result of the person's subjective judgment of his status in the competition. In many competitive and risky situations, including competitive sports, warfare, and securities trading, winners assume a dominant position with respect to the losers, and this appears to be the result of the difference in the status of the winners and losers.

In one experiment, four male graduate students participated in a doubles tennis tournament in which the members of the winning team—the team that won three out of five matches—would win \$100. Blood samples were taken an hour before each match and at hourly interval afterwards for three hours. Because one team won the first three matches, only three were played. Two of the matches were decisive, and during the first hour following these matches, the blood testosterone levels increased in one member of the winning team after both games,. but the blood testosterone level increased in the other member of the winning team after the first match and decreased slightly after the second match. But testosterone levels fell dramatically in both members of the losing teams, as is shown in Figure 8.4a. The outcome of the third match was not determined until the end of the match. Each team had won two of the first four sets, and they were tied five games apiece in the last set. In the tie breaker, each team had five points, when one team lost on a double fault. The testosterone levels of both the winning and losing teams fell during the first hour after the set, as shown in Figure 8.4b.[28]

In a subsequent experiment, male graduate students participated in a lottery in which the prize was $100. Blood samples were taken before and after the drawings. In this experiment, there were no differences in the testosterone levels of the winners and losers, which suggests that testosterone levels are affected only in those contests in which the outcome is due to the participant's own efforts, and in this experiment, the participants had no control over the outcome.[29]

Changes in the blood testosterone levels of professional basketball players correlated with each player's assessment of his contribution to the outcome of the game.[28] Among judo wrestlers, post-match blood testosterone levels were uncorrelated with the outcome of the match but correlated with the wrestler's degree of satisfaction with the outcome.[30]

In addition to physical contests, blood testosterone levels respond in other situations. One study found that blood testosterone levels actually decreased in the winner of a New York chess tournament when the winner knew ahead of time that it would be an easy win for him.[31] High blood testosterone levels were found in male medical students about to receive their medical degrees.[32] Another study of males in different professions found that blood testosterone levels were highest among trial lawyers, a highly competitive profession, and lowest among clergymen, one of the least competitive professions.[33] Still another study found that testosterone levels in males rose during and immediately after intercourse but were only lightly higher after masturbation, suggesting that testosterone levels respond more to the contextual aspects of sex than to its physiology.[34]

Vicarious experiences can also affect testosterone levels, which have been found to increase in fans of winning sports teams and to decrease in those of losing teams. In one study, saliva testosterone levels rose in eight male fans watching a basketball game after their school won. In another study involving 21 males watching a World Cup soccer match on television, the saliva testosterone levels rose in fans of the winning team and fell in the fans of the losing team.[35]

Testosterone levels are not necessarily predictive of the outcome of a competition for rank or status and do not necessarily correlate with rank. In one experiment, blood samples were taken of male rhesus monkeys (*Macadca mulatta*) before and after small groups were formed and the alpha male was determined. There was no correlation between blood testosterone levels and rank prior to the formation of the groups, but after the dominance ranking in the groups was established, there was a tenfold increase in the testosterone levels of the alpha male immediately after he had achieved his alpha status. In long-established monkey groups, however, there was no correlation between testosterone levels and status.[36]

In males, testosterone appears to promote aggressive and competitive behavior by providing the anticipation of a psychological reward for success rather than being a psychological or neurological stimulus of aggressive behavior. It acts in a way similar to the way β-endorphins provide a "runner's high" during and after strenuous exercise such as running or swimming. A motivation for undertaking the strenuous exercise is the anticipation of the "high" at the end of the activity, although the β-endorphins also act as a pain suppressant released into the brain after a threshold of pain is reached. The pleasurable effects seems to be a side effect of the pain suppression. The runner's high may have been an evolutionary adaption that allowed humans to run long distances free of pain in pursuit of game and to outrun predators.

Testosterone in Women. Some females undergo prenatal "masculinization" from the testosterone produced in their ovaries and adrenal glands, although not to the same extent as in males. In females this testosterone is precursor of the female estrogen hormones. Some females are also "masculinized" if they are prenatally exposed to excessive amounts of testosterone. In mice, fetuses are packed closely together in the womb so that if a female fetus lies next to or between two male fetuses, the female fetus is exposed to testosterone generated by her brothers, and such females have been observed to be more aggressive than females who have not been exposed to the male hormones. In humans, most female masculinization occurs if the mother has been

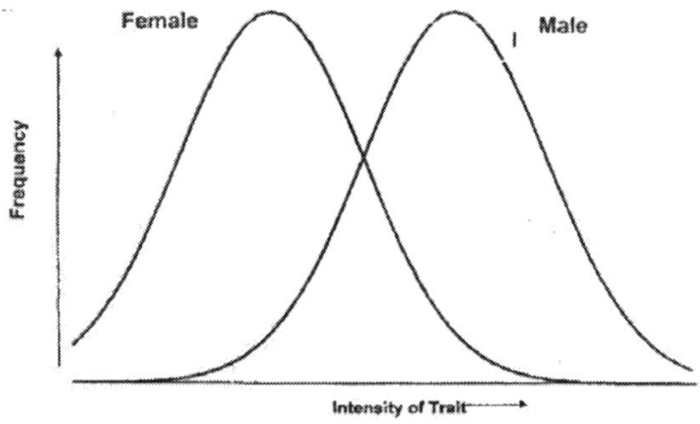

Figure 8.5 - Overlapping Distribution of Male and Female Traits.

treated with synthetic hormones to maintain her pregnancy, or if the fetus was afflicted with congenital hyperpiesia (CAH) which causes the fetal adrenal glands to produce male hormones instead of female estrogens. Girls born with CAH sometimes become transgenic and have sever masculinization of their genitals, display high rates of energetic play, and often initiate fighting and rough and tumble play. They also have a tendency to want to play with boy's toys, and show less interest in infants and dolls than girls normally do.[37]

Although males tend to be more aggressive and dominant than females, the intensity of these traits in both sexes follows normal frequency distributions that overlap as shown in Figure 8.5. The average male is more aggressive than the average female, but some females are more aggressive and dominant than the average male, and some males are less aggressive and dominant than the average female, although the numbers are small. This explains the high aggressiveness of some females and why some are eager to participate in male sports and in combat.

Testosterone and Risk Taking

Testosterone promotes overconfidence and risk taking in men that can result in significant misjudgments and losses. In a recent study, 186 Harvard Business School students—107 men and 79 women—participated in a war game simulation that they played in pairs.[38] The subjects were paid $20 to participate in the game and another $10 if they won. They played on computer terminals but did not know the identity nor gender of their opponents. The players were told to imagine that they were the leader of a fictitious country in conflict with its neighbor over a newly discovered diamond mine located on their disputed border, and that they had $100 million that they could invest in either their military or in their industrial infrastructure, or keep as cash. The player with the most money in her or his industrial account after six rounds won the game. Their strategic options for each round of play were: do nothing, negotiate, make a threat, initiate a military attack against the opponent, continue fighting, or surrender. A military victory depended both on the relative number of battalions that each player had committed to battle and on the outcome of a role of a die. Some games were terminated early because the military forces of both players had been destroyed. The players could negotiate by exchanging messages via their computer terminals. Of the 1080 decisions made by all the players, 70.7 percent were to negotiate, 19.6 percent were to do nothing, 5.9 percent were to fight, 3.7 percent were to make a threat, and none were to surrender. Wars occurred in 47.8 percent of the games.

Saliva testosterone samples were taken when the participants entered the computer laboratory, after three rounds of play, and at the end of the game. Among the females, the mean testosterone levels were 74.9 pmol/l (picomoles* per liter) (s.d. = 53.4), and for males, 341.5 pmol/l (s.d. = 202). Before and after each game the participants were asked to rank their expectations of winning or losing the game on a scale of 1 to 200, with 1 expecting to be the best of all payers and 200 being the worst. Men tended to rate themselves more favorably than women in both pregame and postgame assessments (95 to 78 pregame and 86 and 72 postgame). Postgame testosterone levels correlated positively with pregame self ratings., but postgame testosterone levels and postgame self-self ratings were uncorrelated.

The results of the experiment were: Those players who rated themselves high in their pregame assessments made the most unprovoked attacks on their opponent and the men initiated more unprovoked attacks than the women did, and the men were significantly more likely to attack or retaliate if attacked. Even though the players were unaware of the gender of their opponents, wars were more frequent among male-male dyads, and were least frequent among female-female dyads. The most significant result of the study was the negative correlation between the pregame self-ratings and the outcomes of the games. The most confident players lost more ofte , and this result was greater for the men ($r = -0.16$) than for the women ($r = -0.10$)

---

* A mole is the molecular weight of a substance in grams. Because the molecular weight of water is 18, a mole of water is 18 grams. A picomole is one billionth oof amole so that a picomole of water would 18×10^{-9} grams.

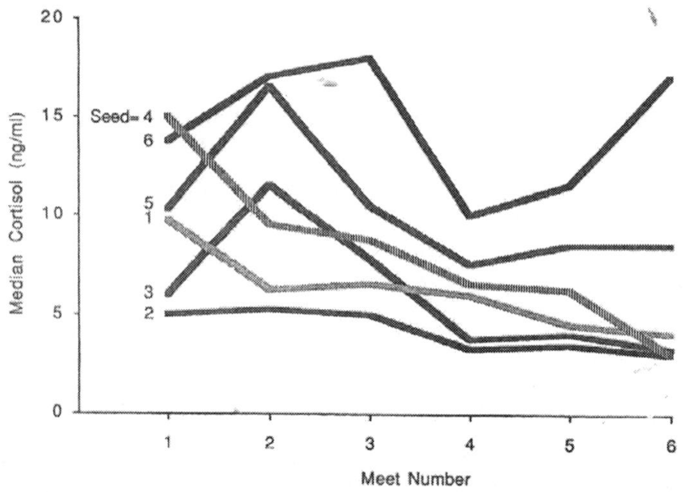

Figure 8.6. Median Cortisone Levels of Six Male Tennis Players.[40]

Another recent study by J.M.Coates and J. Herbert has shown that elevated testosterone levels promote risk taking in male security traders. Saliva samples were taken from 17 male traders on the trading floor of a London bank at 11:00 AM and again at 4:00 PM. On days when the morning testosterone levels of the traders were high, both the average profits and the variance in the profits of those traders were higher than on days when their morning testosterone levels were low. The correlation between the higher variances in their profits and higher morning testosterone levels suggests that testosterone promotes risk taking.[39] Because of the dangers involved in warfare, it reasonable to assume that testosterone in soldiers induces them to take the risks necessary in war.

Cortisol and Dominance

Cortisol is another steroid hormone associated with status in both males and females. Its blood concentrations appear to be correlated with a person's level of anxiety and inversely correlated with a person's assessment of his or her status. The cortisol levels of six members of a men's varsity tennis team tended to fall in the highest ranked players as the tennis season progressed and rose in the lowest ranked players, as show n in Figure 8.6. Cortisol levels were also measured in paratroop trainees over their first 11 days of training. Before their first jump, their average cortisol levels were 87 percent above their basal levels, indicating a high level of anxiety prior to their first jump, but by the 11th day, their cortisol levels had dropped to their pretraining basal levels.[41]

Women and Competition

Women respond differently than men dto the outcomes of athletic competitions and dominance challenges. In volleyball and handball games between elite women athletes, there was no change in their blood testosterone levels or other male hormones during their games. In a study of 17 female rugby players, however, both pregame and postgame testosterone levels were higher than basal levels, but the postgame levels were unrelated to winning or losing, to the player's self-evaluations of their personal performance, or to the player's perceptions of the difficulty of the game. Instead, the rise in pregame testosterone levels were associated with team bonding and being focused on the game but were unrelated to the skill of the opposing team. In contrast, changes in the cortisol levels of the players were positively correlated with the players' evaluation of whether the opponent was more challenging than expected.[35] Another study, however, found a positive correlation between the testosterone levels of 55 women and their occupational status.[42]

The Chemical Mediation of Aggression

Several neurotransmitters are effective in mediating aggressive behavior, including serotonin, oxytocin, and gamma-aminobutyric acid (GABA). The neurotransmitter serotonin, or 5-hydroxytryotamine (5HT), has long been known to inhibit impulsive behavior, including aggression.[43] Lower than normal concentrations of serotonin in the brain have been found in persons with aggressive disorders, including children who display aggressive behavior at home and at school, in military personnel discharged for repeated violent behavior, and in criminals incarcerated for repeated violent crimes. If rats that have not shown aggressiveness toward mice are fed a diet lacking tryptophan, which stops the synthesis of 5HT, they become aggressive, and when they are returned to their normal diet, they revert to their previous state of nonaggressive behavior. Both men and women with acute depletion of tryptophan hydroxylase (HPT) are highly aggressive.[44]

The neurons that produce serotonin are located in the medulla, or brainstem, where the spinal cord enters the brain, and their axons that transmit neural signals extend to many functionally diverse regions of the brain, including the amygdala, hypothalamus, hippocampus, and cerebral cortex.[45] Some human males with impulsive personality disorders are known to have an altered TPH gene that produces HPT that regulates the production of serotonin.[46] Men with the altered form of the gene cannot catalyze the production of serotonin.[47]

On the other hand, monoamine oxidase A (MAO-A) is an enzyme that disposes of excessive amounts of certain neurotransmitters in the brain including serotonin. Men with an altered form of the MAO-A gene causes the synthesis of the enzyme to be stopped half way through its synthesis, causing the altered form of the enzyme to be ineffective in removing the neurotransmitters from the brain. The effect was first discovered in the male members of a Dutch family who exhibited heightened aggressive behavior.[48]

Of the 17 types of serotonin receptors located in various parts of the brain, the $5HT_{1A}$ and $5HT_{1B}$ receptors seem to be those primarily responsible for the mediation of aggression in males.[49] When an agonist* that stimulates either the $5HT_{1A}$ or $5HT_{1B}$ receptor was injected into the hypothalamus of a mouse brain, where the fight-or-flight decisions are made, aggression was reduced without significantly reducing the motor behavior of the mouse, indicating that an aggressive impulse can be suppressed by signals generated in the neocortex where conscious decisions are made and then transmitted to the hypothalamus, where serotonin is released to stimulate the $5HT_{1A}$ and $5HT_{1B}$ receptors in the amygdala. Because the neocortex is the site of conscious decision making, this scenario provides a neuropathway for the conscious mediation of aggressive impulses.[50] In Chapter 10 we discuss how impulsive decisions made in the amygdala are modulated by conscious decisions made in the neocortex.

Oxytocin is a hormone that acts primarily as a neurotransmitter and is implicated in affiliative behavior between individuals and the trust of others. It reduces amygdalae activity in humans and thus suppresses impulsive aggression. Deficits of oxytocin contribute to hostility, fear, and mistrust and thus heighten aggressive impulses. Oxytocin knock-out mice†† display exaggerated aggressive behavior.[51]

---

* An agonist is a chemical agent that acts like a nurotransmitter.

† Oxytocin knock-out mice are mice in which the gene that produced oxytocin has been removed from the fertilized egg.

Most humans are able to control impulsive aggression by signals generated in the frontal cortex that are based on the evaluation of the negative consequences that could result from impulsive aggressive behavior. These signals are sent to the limbic system, where impulsive behavior is controlled. Pathologically excessive impulsive aggression may be due to an imbalance between the drives toward aggression in the hypothalamus and amygdalae, and prefrontal cortex mechanisms that control the aggressive impulses.[52]

Chapter 9

Young Males and Aggression

Now all the youth of England are on fire,
And silken dalliance in the wardrobe lie,
Now thrive the armourers, and honor's thought
Reigns solely in the breast of every man,
They sell their pasture now to buy a horse,
Following the mirror of all Christian kings
With winged heels, as English Mercuris,.
For now sits Expectation in the air,
And hides a sword from hilt unto the point
With crowns imperial, crowns and coronets,
Promis'd to Harry and his followers. . . .

Shakespeare[1]

Since the 1970s, roughly 80 percent of all civil and ethnic armed conflicts occurred in countries with a youth bulge—where the median age was less than 25 years[2]—indicating that the majority of the participants in these conflicts are boys and young men. It is well known that teenage boys take more and greater risks than adult males, and this is likely the reason why many countries and non-state military organizations recruit or draft boys into their armies during their mid-teen years. The enlistment age in many Muslim countries is 16 years, and in many western countries, it is 17 or 18 years. Risk taking reaches its peak in males between the ages between 14 and 25 years, and boys in this age bracket are the greatest risk takers. The reason appears to be that boys in this age group get a greater pleasure reward from their risky activities than older men do. As a person ages, a male's risk-reward calculus changes, so that he gets less of a pleasure reward for the risk activity than a younger males do.

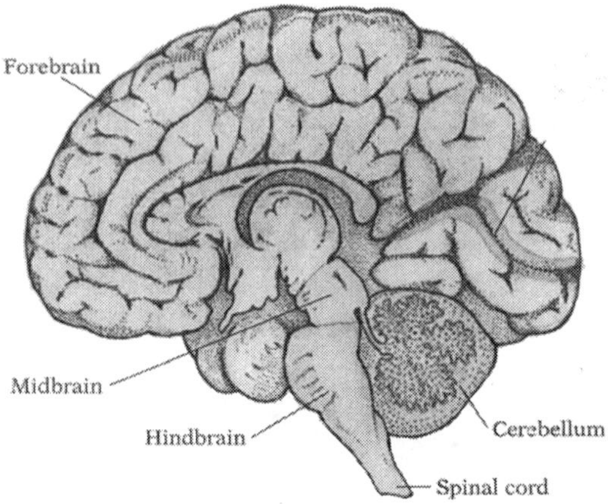

Figure 9.1. The Human Brain[3]

In an experiment, subjects were asked to drive in a simulator through a series of traffic lights in the shortest time. If the light is yellow when they approach it, they have the choice of either stopping or attempting to get through the intersection before the light turns red. If the light turns red before the subject clears the intersection, he is stopped by a traffic policeman, which slows the subject down. Teenage boys are stopped more often than adult males and take longer to run the course than adult males do. Teenagers also take more risks of beating the lights when a friend was present in the simulator.[4]

Why are young males more willing take such risks than older men? To answer this question, we need to understand how the human brain evolved and develops during one's lifetime. The human brain, like that of all vertebrate animals, consists of three major parts, as shown in Figure 9.1: the hindbrain, the midbrain, and the forebrain, the functions of which have become somewhat generalized and overlap. The oldest part of the human brain is the hindbrain, which consists of two parts: the medulla or brainstem and the cerebellum. The hindbrain evolved more than 500 million years ago and is sometimes referred to as the reptile brain. Its main function is to regulate the autonomous functions of the body such as digestion, breathing and heart rate. It also alerts the midbrain and the neocortex—the thinking part of the brain—to the arrival of information from the senses such as visual and audio stimuli, pain, touch, and smell.[5] In animals, the hindbrain is responsible

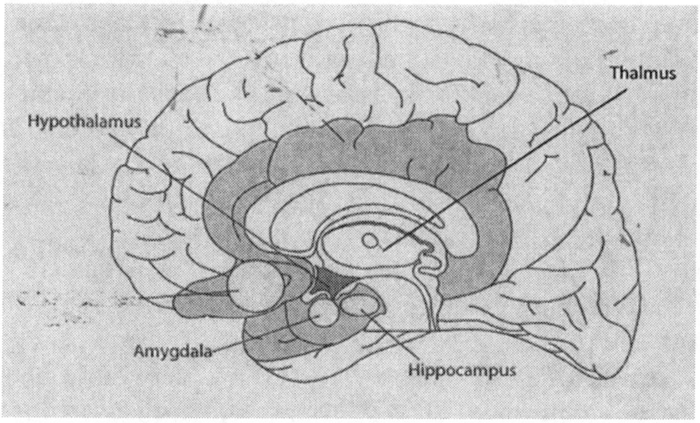

Figure 9.2 – The Limbic System

for primitive functions such as hunting, homing, mating, protecting territory, and fighting.[6]

The cerebellum is wrapped around the medulla and is the "motor" center of the brain, where inherited and certain learned responses such as balance, body position, and motion are stored. Over time, however, the cerebellum has evolved to assume more functions such as controlling body movements through feedback from the rest of the body and relaying information regarding the status of the body to other parts of the brain.

The limbic system, or the midbrain, is located in the center of the brain as shown in Figure 9.2, immediately above the medulla. It evolved between 200 and 300 million years ago in reptiles when its main function was related to the olfactory (smell) sense. The limbic system consists of four parts: the amygdalae, the thalamus, the hypothalamus, and the hippocampus. It is the center of emotions and governs both motivations and behavior. The hippocampus is involved in learning and in long term memory. The limbic system is connected to both the neocortex and to the nucleus accumbens, the brain's pleasure center.

The forebrain is the newest part of the brain. It consists of the cerebral cortex and the cerebrum and is the "thinking part," or "executive branch," of the brain, where judgments and decisions are made using all the information in the brain. The cerebrum is made up of more than ten billion neurons, more than any other part of the brain. Its surface is the cerebral cortex and is where we organize our world and where we make most of our decisions. The cerebral cortex receives information from the body and the outside world and then compares this new information with stored information of prior experiences and knowledge. It then makes decisions and sends instructions to the appropriate muscles and glands through the spinal cord. The cerebral cortex is only about a third of a centimeter thick but is intricately folded, and the human cortex is the most enfolded of all animal cortexes. In humans, the cerebral cortex makes up 70 percent of the entire nervous system and is ten times larger than that of a macaque monkey and a thousand times larger than that of a rat.[7] Although we know that certain activities are centered in the cerebral cortex and that it contains certain kinds of memory, we know very little about how it functions. We don't know how or where memory is stored in the cortex or how memories are retrieved.[8]

The human brain reaches ninety percent of its size by the age of six, but between the ages of 12 an 25 years, the brain undergoes major changes, including a major reorganization. The connections between the brains neurons increase and become more complex, and the axons which send signals from one neuron to another become covered with myelin—the white matter in the brain—that speeds up the transmission of signals between the neurons and consequently the bandwidth of the connections by a factor of as much as a hundred. The dendrites which collect signals from axons develop more branches so that there are many more connections in the brain. The frontal areas of the brain, where decisions are made, become faster with many more connections. Synapse—the interfaces between the axons and dendrites—that are no longer used, wither away—a process called synaptic pruning. The links between the hippocampus and the frontal areas of the brain, where goals are set and different behavioral options are assessed, increase and grow stronger, so that the process of integrating memories of previous experiences into the decision process is greatly improved.[9]

During a person's life, his brain develops in the same order that it evolved, with the hindbrain developing first, then the limbic system, and lastly the forebrain, which usually is not completely developed until a person reaches his mid or late twenties. This means that until the forebrain is completely developed, the limbic system plays a greater role in determining a young person's response to external stimuli and is why adolescents are more prone than mature adults to engage in risky behavior and why young adult males are more likely to become soldiers and suicide bombers than are more mature adult males. Thus, many young males have strong aggressive tendencies and are eager to fight, but are less able to make moral judgments and restrain their aggressiveness. They are easily convinced that they can kill other humans with impunity.

Child Soldiers

Because of their fearlessness and adventurousness, boys as young as ten years have volunteered for or have been involuntarily pressed into military service by both national armies and non-state armed groups. Most countries set a minimum age for enlisting in their armed forces at 17 or 18 years, but some Muslim countries accept boys as young as 16 years, following Quranic tradition. It is estimated that in 2008, there were between 200,000 and 300,000 children under the age of 18 serving as soldiers in the armies of 11 countries, including Burundi, Chad, Ivory Coast, the Democratic Republic of the Congo, Sudan, South Sudan, Bahrain, Afghanistan, Burma and Iran, and in 25 non-state armies.[10]

The most egregious use of child soldiers was by Iran during the Iran-Iraq War in the 1980s. In most armies, child soldiers are used as auxiliaries: observers, orderlies, messengers, and spies. During the Iran-Iraq war, Iran recruited more than 400,000 volunteers, some as old as 70 years, but it is estimated that 300,000 of them were children between the ages of 9 and 18 years who were used as soldiers in that war. The Iranian government has denied the used of boys younger than 16 years when Iranian males come of age and compulsory service for males begins. The boys were recruited by appealing to their patriotism and their desire to become martyrs. The sole requirement was their belief in Allah. Iran's leadership urged boys to take part in the fighting, and the supreme leader of Iran, Ayatollah Khomeini, declared in a series of fatwas issued in 1982 that Allah endorsed their struggle, that volunteering for military duty was a religious obligation, that parental permission was unnecessary for those going to the front and that serving in Iran's armed forces took priority over all other forms of work and study.[12]

Most boys received one to three months of training before being sent to the front, but some boys received no training at all. When they arrived at the front, they were given plastic "keys to Paradise" made in Taiwan to wear around their necks and promised that they would go directly to paradise if they died as martyrs in the war.[13] Many of the boys and some girls were used as human mine sweepers. They were roped together in groups of 20 and then sent across a minefield in front of advancing troops to detonate land mines. One official commented that children were more expendable than regular soldiers.[14] There are

Figure 9.3. Child Soldiers in the Omo Valley, Ethiopia and in the Democratic Republic of the Congo[11]

estimates that as many as 93,000 child soldiers lost their lives during the war.[15]

During the Liberian civil war (1988-2003), mad generals conducted ritual sacrifices of children before battles, and then sent children as young as ten years into the battle naked except for shoes and teddy bear backpacks. The children were fueled with amphetamine and carried M-16 rifles. Some girls wore wedding gowns and Halloween masks and were raped by soldiers [16]

During the American Civil War, thousands of boys under the age of 18 volunteered in hopes of adventure and glory. It is estimated that at least 100,000 Union soldiers were boys under the age of 15 years.[17] Most were regimental drummers, musicians, orderlies, messengers, and observers, but thousands of others fought alongside the regulars. Many of these boys went into battle with only a few weeks of training, but it didn't take them long to understand what they had gotten themselves into. After the Battle of Shiloh in April 1862, Elisha Stockwell Jr. from Alma, Wisconsin, who had enlisted at the age of 15, wrote, "I want to say, as we lay there and the shells were flying over us, my thoughts went back to my home, and I thought what a foolish boy I was to run away and get into what a mess as I was in. I would have been glad to have my father coming after me." Harry Kiefer, a musician with the 150th Pennsylvania regiment, wrote about his experiences at Gettysburg: "[I am called] to look after some poor soldier whose arm is off at the shoulder, and it was just time I got away, too, for immediately a shell plunges into the sod where I had been sitting, tearing my stretcher to shatters."[18]

Violent Video Games and Aggressive Behavior.

The young men who committed the mass murders at Columbine High School, Colorado, in 1999; at the movie theater in Aurora, Colorado, in December 2010; and at Sandy Hook Elementary School in Newtown Connecticut in December 2012 all played violent video games. About 80 percent of American boys between 14 and 18 years play violent video games such as *Call of Duty: Black Ops 3, Killzone 3, Mortal Kombat, Deadly Alliance,* and *Battlefield 3,*[19] which raises the question, whether playing such games is responsible for violent aggressive behavior including murder, rape, and assault. Some research has been done on this subject, but the available data is too sparse to provide definitive answers.

A meta-analysis that integrated the results of 136 separate studies involving 230,396 subjects concluded, "Exposure to violent video games is associated with higher levels of aggressive behavior, aggressive cognition, aggressive affect and psychological arousal and with lower levels of prosocial behavior."[20] A previous meta-analysis that integrated the results of 46 studies of the influence of violent video games on violent behavior concluded, "Exposure to violent video games increases aggressive thoughts, feelings, and behavior, increases arousal, and decreases helping behavior. There was no evidence of moderator effects. That is, these effects appeared to be about the same for males and females, for youths less than 18 years as well as for older participants."[24] Other studies have found that "youngsters who develop a gaming habit can become slightly more aggressive as measured by clashes with peers, for instance—at least over a period of a year or two."[22] Craig A. Anderson, a psychologist at Iowa State University, remarked, "If you look at the literature, I think it is clear that violent media is one factor; it's not the largest factor, but it's also not the smallest."[23] Other researchers have found "that it is aggressive children who are most likely to be drawn to violent video games in the first place: they are self-selected to be in more schoolyard conflicts. And some studies are not able to control for outside factors, like family situation or mood problems."[24] In contrast, Michael Ward, an economist at the University of Texas, Arlington said, "We found that higher rates of violent video game sales relate to a decrease in crimes and especially violent crimes."[25] It may be that playing video games for hours every day keeps people off the streets who would otherwise be getting into trouble. It could be that the games provide an outlet that satisfies violent urges in some players.

Chapter 10

Overconfidence and Decision Making Under Stress

> Reason is, and ought only to be the slave of passion.
>
> David Hume[1]

> Never, never, never believe any war will be smooth and easy, or anyone who embarks on the strange voyage can measure the tides and hurricanes he will face.
>
> Donald Rumsfeld[2]

Decision making in complex situations, especially those involving public and foreign policy, personal and public finances, and matters regarding war, peace, and military operations often result in unexpected outcomes that can be catastrophic. Understanding how individuals and organizations such as governments make decisions and the factors that influence both individual and group decision making may help both individuals and groups such as governments avoid undesirable and catastrophic outcomes. In this chapter we explore how both individuals and groups make decisions, and the factors that can lead to undesirable outcomes.

Decision Making Under Stress

When a person faces a life threatening situation, his initial reaction is likely to be to freeze and then after a few moments he may decide either to run away from the threat or to stay and confront it—the fight or flight syndrome. Because our prehistoric ancestors often faced life threatening situations such as a charging animal or an attacking enemy,they had to decide quickly how to respond to the threat. As a

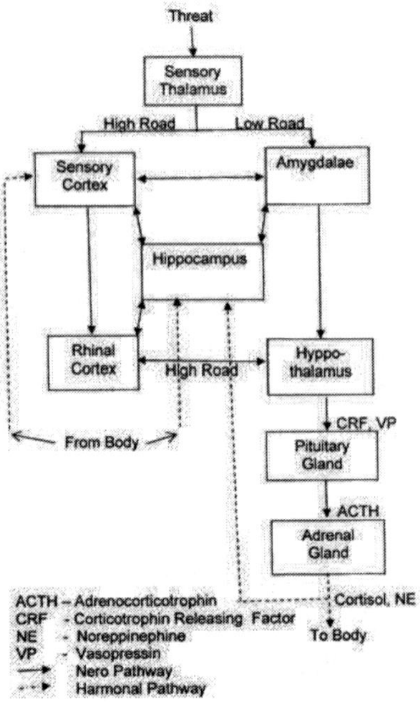

Figure 10.1 - The Threat Response System

result, mechanisms for coping with such situations evolved which we have inherited. But as our ancestors began living in large societies, their lives became more complex as did the threats that confronted them and other mechanisms evolved to cope with these more complex threats.

The quick instantaneous responses described above are made in our limbic system that is locaterd above the hind brain as is shown in Figure 9.2 (page 141). The limbic system is part of tthe autonomic nervous system that controls internal body functions such as heart and respiration rates and digestion. It also contains the amygdalae and the hippocampi that form the threat response system.

The amygdalae are two small organs about the size and shape of almonds which, together with the two hippocampi, are located in each hemisphere of the brain. The hippocampi are involved in short term memory of events and of emotional experiences. However, we are unconcious of activity in our limbic system and thus may be unaware of the motives of some of our actions. In conjunction with the amygdalae, the hippocampi are also involved in fear conditioning in which specific warnings such as load noises or an attacking animal are associated with specific threats and are stored in the hippocampi. The hippocampi also contain inherited fears such as the fears of unfamiliar faces, loud noises and of snbakes and spiders.

When a person experiences a threatening situation, the visual and audio sensations of the event enter the brain through the sensory thalamus through which all sensory information except smell enter the brain. The warnings are sent from the sensory thalmus to the amygdalae where it is evaluated againt conditioned stimuli stored in the hippocampi and a preconditioned response to the threat is generated. The inital response may be to freeze until this process is completed. The decision then may be either to stay and confront the threat or the run from it—the fight or flight syndrom. This pathway is what Joseph LeDoux calls the low road because it provides only a quick and durty rsponse to the threat.[3]

There is a second pathway that can provide a more accurate and judicious response to the threat that is also shown in Figure 9.1 In addition to going to the amygdalae, the threat also enters the sensory cortex which is located in the neocortex—the outer layer of the brain—where the visual and auditory information and information regarding the status of the body are processed. The outpuuts of the sensory cortex are sent to both the rhinal cortex that is also located in the neocortex and to the hippocsmpi.The rhinal cortex receives information regarding the threat and integrates it with information stored in the frontal cortex to create a more accurate representation of the threat anbd an appropraite response to it..

The first pathway which bypasses the sensory cortex and the rhinal cortex is shorter and faster than the second pathway and provides the amygdalae with a quick but less accurate, evaluation of the threat. When the threat is processed through the rhinal cortex, the amygdalae receives a more complete evaluation of the threat, but at a much slower pace because many more of the brain's synaptic connectioins are involved. Joseph LaDoux has called this pathway the high road.[4]

If the amygdal;ae senses a threat, it signals the hypothalamus which rel;eases corticotrophin releasing factor (CRF) and vasopressin (VP) to the pituitary gland that is located at the base of the brain, which then releases ACTH into the adrenal gl;and locatd just above the kidnies which in turn rel;eases hormones into the blood stream to other organs of the body such as the muscels, heart and respatory system which produces a physical response to the threat.

When a country faces a military threat or a call to arms, a majority of its population is likely to respond to the threat in the same way as the Europeans did in August 1914 and as the Americans did after the Japanese attack on Pearl Harbor on December 7, 1941 and after the attack on the World Trade Center and the Pentagon on September 11, 2001. Thus the r*age militarie* described in Chapter 6 arises because all humans possess the same conditioned response mechanisms described above.

Although this system fulfills many of our decision making needs, it is prone to critical limitations. Our long-term memory may not contain information relevant to the threat. The available information may be conflicted, or the problem may be too complex or may contain too many variables and different people will form different responses to the threat. Such situations may lead to extensive debates in which the various sides in the debate draw on different perceptions of the threat or different ideologies may drive the debate.

In many cases ideological predispositions drive a decision making process. During the deliberations leading to the invasion of Iraq in 2003, neoconservatives in the Bush administration believed hat the United States had a moral obligation to establish a western style domocratic government in Iraq. But the attempt to establish such a government by the United Sates was one of the causes of the ethnic and religious conflicts that arose in Iraq following the military victory of the coalition forces.

Overconfidence and Risk Taking

One of the most pervasive human traits is the propensity towards being overconfident in one's abilities—hubris. Although overconfidence can be advanageous in some situations such as courting, job seeking, competitive sports, campaigning for an elective office and investing, it can also resultin poor judgment, errors in decision making, and the underestimation of one's opponents' capabilities. In many decision making situations, overconfidence results in underestimating the risks involved. Nevertheless, a decision to proceed in spite of the risks in many Although this system fulfills many of our decision making needs, it is prone to critical limitations. Our long-term memory may not contain information relevant to the threat. The available information may be conflicted, or the problem may be too complex or may contain too many variables and different people will form different respon ses to the threat. Such situations may lead to extensive debates in which th e various sides in the debate draw on different perceptions of the threat or different ideologies may drive the debate.cases results in a favora ble outcome. For example, some people believe that they can make better investment decisions than the investment professionals, so one deci des to make an investment that a professional investment advisor would c onsider to be too risky for that person's circumstances, yet the investment pays off with a handsome profit. In such situations, the investor was lucky and beat the odds

One of the most common examples of h u b r i s o r over-confidence is the decisions made by national political and military leaders when they decide to go to war. Such leaders usually overestimate their strengths and abilities relative to those of their enemies. In August 1914, the leaders and citizenry of all of the belligerents thought that what turned into World War I would be a short war—that their boys would be home before Christmas—but the war lasted for four years and resulted in sixteen million fatalities and another twenty-five million wounded and missing. In another example, when Japan attacked Pearl Harbor in 1941, Japan's leaders thought that the United states would not fight Japan and that if it did, Japan's Navy would defeat the US Navy. The war in the Pacific lasted 3½ years, and much of Japan was destroyed by the US's conventional bombing and by the two atomic bombs that were dropped on Hiroshima and Nagasaki. I n another example, when President Truman decided to undertake a police action after North Korea invaded South Korea in June 1950, he and his advisors did not anticipate a long war, or that China would enter the war. Finally when the George W. Bush administration decided to attack Iraq in 2003, it was able to defeat Iraq's military forces in 20 days, but it did not anticipate the ethnic and religious wars against the US forces that would follow the initial military victory.

If overconfidence is so common and the cause of so many costly mistakes, how could it have evolved? First, overconfidence may be advantagous beccause it encourages individuals and groups to seek resources that they otherwise would not, and more often than not, they are successful. Even if some losses are costly, over the long term, overconfidence and risk taking have been rewarding. Primitive hunters took great risks in hunting large game, and some of those hunters lost their lives, but the nutritional value of the game was sufficient to justify their losses. To take those risks, those primitive hunters evolved the capacity to underrate them by overrating their hunting abilities. Dominic D.P. Johnson and James H. Fowler have shown that if the value of contested resources is sufficiently greater than the expected costs of obtaining them during their evolutionary history, humans would have evolved a bias toward overconfidence or hubris.[8] They have also shown that "we should expect extreme levels of over confidence (hubris) or under confidence (fear) precisely when we are dealing with unfamiliar or poorly understood strategic contexts. . . . We predict that where the value of a prize sufficiently exceeds the cost of competing, overconfidence will be particularly prevalent in very important domains that have inherently high levels of uncertainty."[9]

War is a contingent strategy that confronts a society and its leaders with policy choices. Not every hostile encounter between human societies results in war. Societies, like individuals, are faced with the fight-or-flight syndrome, and after an initial decision to respond aggressively, the society may decide not to fight. Even among very hostile societies, fighting is frequently dormant for long periods because of such factors as the weather, unpreparedness or war weariness.

For every species there appears to be an optimum level of aggressiveness that maximizes its survival because of the time and energy spent in aggressive activities is time and energy that could be spent in more productive activities. This is also true for human societies, where the level appears to be relatively high. Some ethnic groups such as Israelis and Palestinians, Catholics and Protestants in Northern Ireland, have engaged in low-level warfare for many years, at levels that appear to be acceptable to both sides. To settle such disputes may be thought to be more costly or difficult for one or both sides than to continue fighting, or the parties may not be able to reach an agreement to end the conflict. Such conflicts have been termed frozen conflicts because they cannot be resolved and a low level conflict continues.

War Risks and the Value of Human Life

When a country is at war, it is willing to commit extraordinary resources, including its young men, to achieve victory or to avoid defeat. During World War I Germany France and England threw hundreds of thousandss of young men and boys into battle after battle on the western front where they were slaughtered like sheep. More than a million lives were lost in the first battle of the Somme in August 1914, and the British lost more than 60,000 men on the first day of the Battle of Passendale on July 1, 1916. For the generals who committed those soldiers into those battles and the statesmen who agreed to continue the war, the value of the lives of these humans must have been very low, because they were willing to sacrifice so many young men, but even the generals despaired of the war. In November 1917 during the final stages of the British Passendale offensive, in which the English had sacrificed more than 400,000 of their young men, Sir William Robertson, Chief of the Imperial Staff, wrote to British Field Commander Sir Douglas Haig, "I confess I stick to it more because I see nothing better, and because my instinct prompts me to stick to it, than because of any good argument by which I can support it," to which Winston Churchill commented, "These are terrible words when used to sustain the sacrifices of nearly four hundred thousand men."[11]. Robertson and Haig were faced with a situation they didn't know how to deal with and allowed hundreds of thousands of English youth to be killed.

On the western front, World War I devolved into a trench war after the initial German offensive in which an offense consisted of a mass of soldiers rising out of their trenches and advancing across an open no man's land into the fire of an entrenched enemy armed with machine guns and artillery. It amounted to a slaughter, and little or no territory was gained, yet the lives of hundreds of thousands of young men were sacrificed. It has been said that Europe lost a generation of its manhood during that war. To the generals and political leaders, the value of an infantry soldier's life must have been negligible compared to their war objectives.

The terrible death tolls of the war disgusted many of the junior officers, especially German officers, which motivated them to develop the Blitzkrieg, or mechanized warfare, in which tanks, infantry, artillery, and combat engineers were deployed in highly mobile units that moved rapidly against the enemy. Although the tank had been introduced during World War I, it was developed into an effective effective weapon, together with the armored personnel carrier and mechanized artillery, during the interwar period. By 1939, all the western armies had become mechanized, and the German army used the Blitzkrieg in their offensives against Poland in 1939, against France and the low countries in 1940, and against Russia in 1941. The United States and the British employed mechanized units in the final stag es of the war to defeat Germany. The result was a tremendous decrease in combat casualties, but because of the strategic bombing campaigns of the war, civilian casualties, which, except on the eastern front, were minimal in th e First World War, became a major factor in the Second World War. These changes in strategy reduced British combat fatalities from 10 percent in World War I to 5.2 percent in World War II, but in spite of the German Army's adoption of the Blitzkrieg, German Army fatalities increased from 18.6 percent in the First World War to 28 p ercent in the second war, sustained mainly on the Russian front in the Battle of Stalingrad and the retreat of the German Army from Russia d uring the winter of 1944–1945.

A Case Study: Operation Iraqi Freedom

The reasons and motivations for the US invasion of Iraq in March 2003 remain unclear to this day. The George W. Bush administration claimed that it invaded Iraq to destroy its weapons of mass destruction (WMD), which it alleged that Iraq possessed· High-level members of the Bush administration, including Vice President Richard Cheney, Secretary of Defense Donald Rumsfeld, Deputy Secretary of Defense Paul Wolfowitz, the President's N ational Security Advisor, Condoleezza Rice, and National Security Advisor to the Vice President I. Lewis (Scooter) Libby, and President Bush himself, all seemed anxious for the United States to invade Iraq. In addition to removing Iraq's alleged WMD, several other reasons have been alleged for the invasion, including: revenge for assassination attempts on the president's father in 1993 during his trip to the Middle East; the war hawks in the administration wanting their administration to be a war administration; President Bush himself wanting to be a war president; and a widespread belief of the neoconservatives within the administration that freedom and democracy are universal human aspirations, not just American ones, and the United States had a moral obligation to establish democratic governments throughout the world, including in the Muslim world, using its military force if necessary.

As early as August 1, 2001, the Deputy's Committee* issued Top Secret paper entitled "A Liberation Strategy" that proposed a strategy for pressuring Saddam Hussein to give up his WMD, but it did not suggest a regime change in Iraq. The paper was devoted a mostly to diplomatic and covert measures, and it promoted a discussion among the principles and their deputies about the circumstances under which the US miliary might use to achieve these goals.[12]

The first discussion of going to war in Iraq occured in meeting of the National Security Council (NSC) on September 12, 2001, the day after the terrorist attack on the World Trade Center in New York and the Pentagon in Washington. During a discussion of how to attack al Qaeda, Secretary of Defense Donald Rumsfeld asked President Bush, "Why shouldn't we go after Iraq, not just al Qaeda?"[13]

On September 16, 2001, five days after the terrorist attacks in New York and Washington, President Bush told Condoleezza Rice, his national security advisor, "We won't do Iraq now, we're putting Iraq off, but eventually we'll have to return to that question." Then on November 21, 2001, Bush told Rumsfeld that "it was time to turn to Iraq," and he wanted to know what it would take to protect America by removing Saddam Hussein if we had to.[14] Rumsfeld immediately told General Tommy Franks, the commander of Centcom (the Central Command) to review the Pentagon's war plans for Iraq.

On October 19, 2002, Rice circulated the National Security Presidential Directive (NSPD) "Iraq: Goals, Objectives and Strategy" to the principles. It stated that the US goals were to create an Iraq that: "Does not threaten its neighbors; Renounces support for, and sponsorship of, international terrorism; . . . Is free of weapons of mass destruction (WMD), their means of delivery, and associated programs; . . ." The US objectives were to "To achieve this goal in a way that: Minimizes the risk of a WMD attack against the United States, U.S. field forces, our allies and friends; Does not contribute to regional instability; . . . Ends Iraq as a safe haven for terrorists." The strategy included: "Employ all insturments of U.S. national power in a coordinated fashion; Pursue our goals our goals and objectivees with a

---

* The principles included the vice president, the president's national security advisor, the Secretaries of State and Defense, and the director of the CIA. The deputies included the deputy secretaries of the State and Defense Departments, the deputy director of the CIA, the deputy national security advisor, and the vice president's national security advisor.

coalition of committed countries; Secure[s] UNSC (United Nations Security Council) approval if possible; . . . [Establishes] an interim administration in Iraq that prepares for the transition to an elected Iraqi government as quickly as possible; . . . [T]ransitions to an elected Iraqi government."[15]

The Presidential Directive did not rule out the use of military force or an invasion of Iraq, and indeed, the decision to invade Iraq may have b een made in October 2002 when the directive was issued. On October 16, 2002, the US House of Representatives voted 297 to 133 in favor of a resolution to use military force against Iraq, and the Se nate voted 77 to 23 in favor of the House Resolution. On November 8, 2 002, at the behest of the United States, the UN Security Council passed UN Security Council 1441 that said that if Saddam Husain contin ued to violate his disarmament obligations he would face "serious conse quences," a phrase that replaced the US preferred wording, "all necessary means." The resolution was passed 15 to 0. However on January 31, 2003, Bush meet with Tony Blair, the British prime minister in the Oval Office, Blair told Bush that he had to get a second UN resolu tion, because he had promised it to his political party and it was politicall y necessary for Blair.

The best chance for obtaining a second resolution authorizing the use of force against Iraq was to convince the international communit that Iraq was in violation of the UN sanctions against its possession of WMD. The Security Council had lready passed ten resolutions that The Security Council had already passed ten resolutions that condemned Iraq for its possession of WMD and for its interference with the UN inspections during the previous 12 years.

On Saturday morning, December 21, George Tenet and John McLaughlin, CIA director and deputy director, briefed the president in the Oval Office on the intelligence concerning Iraq's WMD. Using flip charts, McLaughlin told the president and his advisors that critical components of the biological weapons could not be accounted for, including 3,200 tons of precursor chemicals and some 6,000 shells to deliver the agents. He displayed a satellite photograph of a chemical weapons facility that showed scarre d earth that appeared to be left after an attempt to clean up after a spill or transfer of chemicals. Another photograph showed an unmanned aerial vehicle (UAV) flying in a circular flight pattern that McLaughlin said had flown a total of 00 kilometers, a clear violation of a UN resolution that limited the range of Iraq's UAVs to 150 kilometers. McLaughlin then presented accounts of two intercepted conversation between two Republican Guard officrts discussing the removal of "nerve agents" from the facility and the other intercept in which he claimed that the officers were discussing the concealment of a modified vehicle at the al-Kindi company, a known WMD facility, because inspectors were scheduled to arrive there.[16]

After Mc Laughlin finished, Bush said, "Nice try," and then turned to Tenet and said, "I've been told about all this intelligence about WMD and this is the best we've got?" Tenet then rose from one of the couches in the Oval Office, threw his arms in the air, and said, "It's a slam dunk case." Bush asked, "George, how confident are you?" Tenet replied, "Don't worry, it's a slam dunk." Later, Bush recalled that McLaughlin's presentation "wouldn't stand the test of time" but that Tenet's reassurance "was very important." He told Chief of Staff Andrew Card and Rice, "Needs a lot more work," and several times, he told Tenet, "Make sure no one stretches to make our case."[17]

Rationalizing the Decision. The task of obtaining the UN resolution that Tony Blair had told Bush he needed fell to Secretary of State Colin Powell. To prepare himself for a Security Council briefing, Powell reviewed the available intelligence on Iraq's WMD but found it murky and unconvincing. Iraq had used chemical weapons in its war with Iran during the 1980s and against Iraq's Kurdish civilians in northern Iraq. The Iraqis had hidden them from the UN inspectors in the 1990s, but there was no conclusive intelligence indicating that he still possessed the weapons.* After the 1991 Gulf War, the UN had put sanctions on Iraq and had embargoed Iraqi oil exports depriving Iraq of $100 billion annually in oil revenues. Why would Iraq suffer these sanctions and forgo the oil revenues for all those years? Why didn't Saddam Hussein allow the weapons inspectors in and come clean if Iraq didn't still have the weapons? But the intelligence remained murky.

Powell was sechuled to brief the UN Security Council on February 5 so on February 1 he went to the CIA headquaters in McLean Virgina to review the raw intelligence availabel concerning Iraq's WMD. He spect that day and evening and the next day at the CIA, but there was no conclusive evidence that Iraq still had any of the banned weapona.* Powell finally found a communications intercept

*In 2004 the US Army personnel began finding large caches of buried 155mm artillery shells and 122mm rockets filled with muster gas and sarin nerve gas that had been buried before the U.S. began its air attacks on Iraq in 1991.

from the day before the inspectionnnns had begin the previous November in which a colonel told a brigadier general that he had modified a ehicle that had been used in the WMD programs in the past. The colonel then said, "We have evacuated everything. We don't have anything left." Another intercept referred to "forbidden ammo," and in a third, a colonel instructed a captain to remove the term "nerve agents" from wireless instructions.[18] Although these intercepts may have been incriminating, they did not provide the unequivocal evidence that Powell s ought.

When Powell briefed the Security Council on February 5, 2003, he told it, "I cannot tell you everything that we know. But what I can share with you when combined with what all of us have learned over the years, is deeply troubling. What you will see is an accumulation of facts and disturbing patterns of behavior. The facts on Iraq's behavior demonstrate that Saddam Hussein and his regime have made no effort—no effort —to disarm as required by the international community." He then played thethree intercepts, to which he added his personal interpretations. Concerning the "forbidden ammo," he gratuitously added, "Clean out all of the areas, the scrap areas, the forbidden areas. Make sure there is nothing there," which was not in the intercept. He told the Security Council, "We know Saddam Hussein has what is called 'a higher committee for monitoring the inspection teams. . .Think about that. Iraq has a high-level committee to monitor the inspectors who were sent there to monitor Iraq's disarmament. Not to cooperate with them, not to assist them, but to spy on them and keep them from doing their job assist them, but to spy on them and keep them from doing their job. . . . We know from sources, that a missile brigade outside Baghdad was dispersing rocket launchers and warheads containing biological warfare agents to various locations." Powell cited satellite photographs and other intelligence that suggested that the Iraqis were cleaning up old chemical and biological weapons facilities prior to the arival of the UN inspectors. He then showed a slide of what he said were biological weapons factories on wheels which he claimed were based on several human sources.[19] Powell concluded his briefing by saying, "We know that Suddam Hussein is determined to keep his WMD, .he's determined to make more. . . .[S]ould we take the risk that that he will not someday use these weapons at a time and a place and in a manner of his choosing, at a time when the world in a much weaker position to respond. The United Stats will not and cannot run that risk to the American people."[20]

The human source cited by Powell was Rafid Ahmad al-Janabi, codenamed Curveball, who was a former Iraqi citizen who claimed to be a chemical engineer who had worked in Iraq's mobile chemical weapons laboratories. In 1999 al Janabi defected to West Germany where he applied for political asylum and then made his claims in an attempt to obtain a US green card. The Germans did not allow US intelligence agents to question Curveball but passed his statements on to the US Defense Intelligence Agency. He gave many hours of testimony about Iraq's WMD programs and about the mobile laboratories in particular, but his veracity was questioned by both the German Federal Intelligence Group and by the British Secret Intelligence Service. He was later found by the Iraqi Survey Group to have lied.

In his 2003 State of the Union address, President Bush stated, "From three Iraqi defectors we know that Iraq, in the late 1990s, had several mobile biological weapons labs. These are designed to produce germ warfare agents, and can be moved from place to place to evade inspectors. Saddam Hussein has not disclosed these facilities. He's given no evidence that he has destroyed them."[21]

The United States did not introduce a second resolution in the Security Council, because it knew that it would not have enough votes to pass the resolution. The presidents of Mexico and Chile refused to support a resolution. On February 7, Bush talked to the French president, Jacques Chirac, who told Bush, "War is not inevitable. There are alternate ways to reach goals. It's a question of morality. I'm against war unless it is inevitable and necessary. . . . I fully understand your position is different. There are two different moral approaches to the world, and I respect yours."[22] Washington seemed confident that France would abstain and would not veto a new resolution, but on February 20, Chirac, Russian President Vladimir Putin, and Gerhard Schroeder, the German chancellor, issued a joint statement calling for extended WMD inspections. Chirac said, "Russia, Germany and France are determined to ensure that everything possible is done to disarm Iraq peacefully."[27] The administration now knew that it would be very difficult to pass a new resolution in the Security Council and instead decided to use the previous Resolution 1441, which said, "The Council has repeatedly warned Iraq that it will face serious consequences as a result of its continued violations of its [disarmament] obligations," to justify an attack on Iraq.

The Invasion and Its Aftermath. On March 20, 2003, a coalition of 248,000 US, 45,000 UK, 2,000 Australian, and 198 Polish troops invaded Iraq from Saudi Arabia. Baghdad fell on April 9, twenty days after the invasion. On May 1, 2003, President Bush donned a Navy flight jacket and boarded the aircraft carrier *USS Abraham Lincoln,* just off the coast of San Diego, and addressed the country from its flight deck under a huge bannr that read,**MISSION ACCOMPLISHED.** He declared, “Major combat actions in Iraq have ended. The tyrant has fallen and Iraq is free.” He said that it was a “noble cause" and "a great moral advance.”

Immediately after the defeat of the Iraqi Army, the allied coalition established the Coalition Provisional Authority (CPA) to govern Iraq until a provisional government could be formed. The CPA invested itself with executive, legislative, and judicial authority in Iraq. The CPA was initially headed by Jay Garner, a former US Army lieutenant general and the president of SYColman Company, a defense contractor. Garner had been selected to lead the CPA because of his role in Operation Provide Comfort after the 1991 . Gulf War for the resettlement Iraq's Kurds into their previous homes after they tried to flee Iraq, but hundreds of thousands of them were trapped on a barren rocky hillside on the Turkish border after Turkey had refused them refuge.

Garner was replaced on May 11 by Ambassador L. Paul Bremer as the US administrator of Iraq. Bremer immediately issued two orders: CPA Order No. 1 on May 16 banned all but the lowest members of Saddam Hussein’s Ba’ath Party from the government. A week later, he issued CPA Order Number 2 to disb and the Ministries of Defense and Interior and the entire Iraqi Army. The Ba’ath Party had been Saddam Hussein’s ruling party, composed mostly of Sunni Muslims. The first order threw 50,000 civil servants out of work most of whom had joined the Ba’athe Party only because party membership was required of all government workers. Bremer justified. his actions by rel;easing a statement “to show the Iraqi people that the Saddam regime was gone and will never return.” The order to disband the army demobilized hundreds of thousands of men who possessed weapons and knew where more were stored. Many of the officers a nd enlisted men joined terrorist groups such as Al Qaida in Iraq and later made up the core of the Islamic State insurgency.General David Petraeus later said that "if Beermer's orders didn't create the insurgency that followed, they certainly inflamed and accelerated its rise."[24]

Iraq's population included Muslim Shiites (63–70 percent), Sunni Arabs (15–17 percent) and Kurds (15–20 percent). The remainder of the population—about 3 percent—was composed of various Christian sects.

As soon as the invasion had ended, the coalition forces encountered attacks from the Fedayeen (those who sacrifice themselves), who were Saddam and Ba'ath Party loyalists, who were soon joined by religious radicals and other Iraqis who opposed the coalition's occupation. The insurgents did not welcome the coalition forces as liberators, but regarded them as occupiers and called them crusaders. They employed mortars, suicide attacks, snipers, car bombs, and improvised explosive devices (IEDs) against the coalition forces and sabotaged Iraq's petroleum, water, and electrical infrastructures. After a lull during the winter of 2003–2004, during which the insurgents studied the tactics of the coalition forces and reorganized itself, the attacks intensified in the spring of 2004, and the insurgents were joined by other militants from the Middle East. In Mosel, at least 40 retired Iraqi army generals and hundreds of former soldiers joined the insurgency.[25]

On March 31, the insurgents in Fallujah Province ambushed a convoy of private contractors who were providing security for a food service contractor. Four of the security contractor's men were killed by grenades and small-arms fire. Their bodies were dragged through the streets by their vehicles, then beaten, set afire, and hung from a bridge over the Euphrates River. Both Rumsfeld and Bremer refused to call the war an insurgency, and both prohibited even the utterance of the word in their offices. Rumsfeld knew that if the war was an insurgency, it would require a massive ground force and would take years to defeat at a great cost. Rumsfeld didn't want to get bogged down in securing and stabilizing Iraq after defeating its army, so he hadn't made any plans to fight an insurgency, and he said he wouldn't approve any other plans to do so.[26] Instead the 18,000 man 101st Airborne Division was replaced with a single infantry division of just 9,000 men.[26]

In early 2003, Maqtada al-Sadr, a young Shiite cleric and the son of a prominent Shiite leader, had risen in prominence as a Shiite leader in Sadr City, a Shiite district in Bagdad. Al-Sadr had become a popular charismatic leader who had a lage following in Iraq that included both Shiites and Sunnis. Al-Sadr opposed the occupation and the CPA, and claimed that he had more legitimacy than the CPA, which had been appointed by the invaders and occupiers. In his sermons and public statements, he demanded that the coalition withdraw from Iraq. He organized thousands of his followers into a political movement that included an armed contingent known as the Mahdi Army. In March 2004, the US Army shut down his newspaper, *al-Hawza*, claiming that it was inciting violence. On April 4, fighting broke out in Basra, Sadr City, and Najaf in which dozens of civilians were killed. It was the most serious challenge to the coalition up to that time. In April, al-Sadr sent a convoy of relief supplies to Sunnis in the besieged city of Falluja. In August 2005, several thousand of al-Sadr supporters rallied in support of his ideas.

On January 31, 2005, the Iraqi people elected the Iraqi Transitional Government to draft a new constitution for Iraq, which was ratified in a referendum in November, and the Iraqi National Assembly was elected in December. Although some violence and a Sunni boycott marred the election, most Shiites and Kurds participated in the election. Al-Sadr controlled the largest bloc in the parliament, and after it had selected Nouri al-Maliki as the prime minister, al-Sadr reached a modus operandi with al-Maliki, who depended on al-Sadr's support to retain his position. But al-Maliki's government was composed mostly of Shiites and al-Maliki made no attempt to form an inclusive government that included Sunnis and Kurds, who oppose the government and intensified its insurgency against the government.

After the 2005 parliamentary elections, the CIA interviewed four potential candidates for the Prime Minister, and selected al-Maliki as someone who was "independent of Iran." But he was also supported by the leader of Iran's Quds Force, a unit of Iran's Revolutionary Guards that is under the direct command of Iran's supreme leader, and he also had the support of al-Sadr.

During Salaam Hussein's reign, Nouri al-Maliki was a member of the Shiite Islamic Dawa Party and an anti-Hussein dissident. In July 1979 he went abroad under a death threat and spent the next twenty four years in Iran and Syria. where he became a senior leader of the Dawa Party from which he coordinates the activities of anti-Hussein guerillas in Iraq. He returned to Iraq in 2003 after the US led invasion.

In 2007 al-Sadr went to Iran where he delivered a fiery apeech in which he condemed th US occupation and called for a mass demonstration opposing the occupation, for the Iraqi army to stop cooperating with the American Army, and for the gorilla forces to push the US forces out of Iraq. His statement was distributed in the holy city of Fallujah on March 8, the day before a large demonstration there. Then on April 17, several government ministers loyal to al-Sadr resigned. On May 25, al-Sadr delivered a speech to 6,000 of his supporters in which he again condemned the occupation. In June, he led his followers on a march to the heavily damaged Ashariyya shrine in central Iraq. He said that the purpose of the march was to bring Shiites and Sunnis closer together.

The bloodiest battle of the postwar period was the Second Battle of Fallujah in November 2004. Coalition forces prevailed after 45 days at a cost of 95 US and 1,350 insurgents' lives. It was the bloodiest urban battle since the Vietnam War. The US Army was chastised for using incendiary phosphorous against the insurgents as a violation of the laws of war.

In 2004, the abuse of prisoners at Abu Ghraib prison by the 320th Military Police Battalion of the US Army National Guard was revealed. Pictures taken by the military guards at the prison were released and widely distributed in the press, showing rape, sodomy, and torture of the prisoners. Seventeen officers and enlisted men were relieved of duty, and 11 enlisted persons were court-martialed, convicted, sentenced to military prison terms of six months to ten years, and dishonorably discharge.

In February 2005, the tours of 15,000 US troops were extended to provide security, but May became the bloodiest month since the invasion, when fundamentalist Sunnis from Syria and Saudi Arabia undertook suicide missions against the Shiite population, killing more than 700 civilians and 79 US soldiers. The sectarian violence increased in January and February 2006 and was marked by the bombing of the al-Askari mosque in the city of Samarra, one of the holiest Shiite sites in Iraq and al Qaeda in Iraq was blamed for the bombing. Although the bombing caused no casualties, violence erupted in the days following the bombing, and a hundred bodies were found riddled with bullets. The US Army estimated that the average homicide rate in Baghdad had increased from 11 to 33 per day during January and February. The UN described the situation in Iraq as "civil war like." On March 6, five US soldiers raped a 14-year-old girl, then murdered her, her mother and father, and her six year old sister. They set fire to the girl's body to cover up their crime. The soldiers were court-martialed; four were convicted of rape and murder, and the fifth of lesser crimes. This crime became known as the Mahmudiyah killings.

The 2007 Troop Surge and a Change in Strategy In response to the mounting toll of the insurgency in Iraq, President Bush announced a troop surge of 20,500 troops in five brigades to Iraq on January 10, 2007., and extended the tours of most army troops and of many marines already in Iraq. He also announced that the overall objective of the surge was to establish a "unified democratic federal Iraq that can govern, defend and sustain itself and is an ally in the War on Terror."

In February, President Bush appointed General David Petraeus as the commander of the Multinational Task Force Iraq. Petraeus was a four star general with a PhD in international relations from Princeton University. He also played a major role in the preparation of the new Field Manual on Counterinsurgency Operations (FM 3-24) in which the focus of counterinsurgency operations was not to engage the insurgents directly, but to gain the "hearts and minds" of the population and to protect it from the insurgents through building relations with it and hiring former enemies. Petraeus .has been quoted as saying, "money is ammunition," and he distributed millions of dollars in cash to Sunni insurgent groups in an attempt to buy their loyalty and to protect the population.

But the new strategy didn't accomplish its goals. Fatalities in the population rose by 15 percent from February to March and fell by 18 percent during the same period among the insurgents. By September, civilian causalities began to fall, but only because Maqtada al-Sadr had ordered a six months stand down of his Shiite troops, because of the mounting toll of the Shiite population in Iraq. On May 2007, 144 members of the Iraqi Parliament (52 percent) signed a petition that called on the United States to set a timetable for withdrawing from Iraq, and in June, the Parliament voted 144 to 131 for the US government to consult with it before requesting the UN Security Council to extend its mandate for coalition operations in Iraq that would expire at the end of 2008. Pressures on the coalition forces were compounded in January 2008 when Tony Blair announced that the British would withdraw its forces from Basra and turn over security operations there to the Iraqis. In July, Denmark withdrew 441 of its troops, leaving only nine soldiers to man four observation helicopters.

On December 4, 2008, a month after Barach Obama was elected as the US president, the US and Iraqi governments signed a new status of forces agreement that specified that the US forces would begin withdrawing from Iraqi cities in June 2009, and would completely withdraw from Iraq by the end of 2011, and that US contractors working for the US Army would become subject to Iraqi criminal law. The last US combat brigades left Iraq on August 19, 2010, leaving only 50,000 troops in Iraq to provide support for the Iraqi military.

Post-withdrawal Iraq. As soon as the US and other coalition forces left Iraq in December 2011, a fierce string of terrorist attacks swept across the country, adding a new level of violence to the political and sectarian feuds in Iraq, and threatening to undo the modicum of stability that had existed before the withdrawal. Al Qaeda in Iraq increased its terrorist activities during the month after the US withdrawal and was responsible for more than 60 deaths over a period of three days. The Iraqi security forces seemed to be incapable of dealing with al Qaeda.

On December 26, 2011, Maqtada al-Sadr called for parliament to be dissolved and for new elections. Al-Maliki then announced that his government would welcome the Iranian-backed militia, Asaib Ahl al-Haq (the League of the Righteous), into Iraq's political system. The Asaib militia was one of the deadliest insurgent groups operating in Iraq. It had bombed American facilities and convoys, had assassinated Iraqi officials, and had attacked Americans even as they were leaving Iraq. It was responsible for the last American death in Iraq. Between November 2011 and March 2012, there were almost 400 fatalities resulting from sectarian violence in Iraq.[27]

Just days after the US withdrawal, the Sunni party boycotted the parliament and cabinet meetings in protest of what it saw as attempts by the Shiite-dominated government to marginalize them and push them out of the government. But in February 2012, the Sunnis agreed to return to parliament, paving the way for a national summit meeting to discuss reconciliation between the Sunnis,. Shiites and Kurds, but the summit failed to resolve any of the Sunni and Kurdish complaints against al-Maliki's Shiite dominated govern-ment.[28,29], In January the Human Rights Watch, an advocacy group·, said that the Americans had left "a bidding police state" behind and that the Shitte leadership was ruling Iraq by force and fear that restricted freedom of expressionand and that security forces had intimidated, threatened and detained activists, demonstrators and journalists.[30] At the time of the withdrawal President Obama said that the although Iraqis could still unite, the country was far from "sovereign, stable and self-reliant." From January 2012 through January 2014 terrorist attacks in Iraq left more than 15,000 dead and many more injured.[31]

In January 2012, the Sunni vice president, Tariq al-Hashimi, fled to the semi-autonomous Kurdistan region in the north to escape arrest after the al-Maliki government issued a warrant for his arrest. He was accused of running a sectarian death squad that had assassinated security and government officials. A panel of Iraqi judges said that the death squads commanded by Hashimi had conducted 150 attacks against religious pilgrims, security forces, and its political foes during the previous six years.[32] Hashimi denied the charges and accused al-Maliki of using Iraq's security forces to persecute his political opponents, specifically Sunnis. Hashimi said that he would stay in Kurdistan beyond the reach of the Iraqi police. The next day, al-Maliki threatened to abandon the coalition government that the Americans had left in Iraq.

* * * * *

The United States has paid an incredibly high price for its Iraqi debacle: the lives of 4,498 of its soldiers, sailors and airmen, and more than 32,000 wounded, some of whom are missing limbs that will never be fully replaced, and others with post traumatic stress disorders from which they may never recover. The monetary cost of the war has been estimated to be more than $ 1.2 trillion. The United States left behind a Shiite dominated theocratic state that has been unable to govern its multiethnic population. Major violence broke out again in Iraq in 2014 with rise of the Islamic State that was able to take and hold territory in Syria and northern Iraq.

Among the causes of America's debacle was the almost complete lack knowledge and understanding of Iraq and the Muslim world and its culture by the Bush administration. It didn't understand the animosities that existed among the Iraq's religious and ethnic groups that also exist throughout the Muslim world, and that caused both the Shiites and Sunnis to rebel against the occupation and then for the Sunnis and Kurds to rebel against the Shiite al-Maliki government set up by the American occupation that didn't anticipate that the Iraqis would view the invaders as conquers ansd occupiers and not as liberators and that the

Iraqi's would put up a fierce resistance to the occupation. A major cause of the insurgency was the disbandment of the Iraqi Army that prompted many of its officers and enlisted men to join the insurgency that opposed the occupation and joined al Qaeda in Iraq and later formed the core of the Islamic State that created a caliphate based on the most extreem form of the Shariah Law.

The decision to invade Iraq is an example of using the "high road" to rationalize President Bush's decision made on the "low road" because of his hubris and because his advisors didn't have sufficient information to make an accurate assessment of the situation. Although the United States and its allies won the military battle, they lost the peace and therefore they lost the war in Iraq.

The United States also bears a heavy responsibility for the existence of the Islamic State. First because of its disbanding of the Iraqi Army after the invasion, and secondly, by not insisting that al-Maliki form a truly coalition government that included Sunnis and Kurds in responsible positions.

Chapter 11

Reconciliation

Make peace, make peace, don't fight; if you fight
I'll bite, and we can't bite since we're friends.

Mirilka: A Russian children's
peacemaking rhyme

Every war ends, but few end with the complete annihilation of the defeatted such as the destruction of Carthage at the end of the Third Punic War in 146 BCE. Some wars end in recrimination and bitterness that are the genesis of a new war as soon as the antagonists can recoup and rearm. Others end with a reconcilation betweeb the antagonists in which they are able to arrange a viable and lasting peace. World War I is an example of the first, and World War II is an example of the second. In this chapter we investigate some of the techniques that have been used for reconcilation between antagonistic groups. These include third-party mediation, and superordinate tasks that benefit all the parties, such as the unification of Europe after World War II, but this approach can be successful only if all the former antagonists participate.

Reconciliation among Primates

In chimpanzee communities, males fight one another to achieve dominance, and such fights usually take the form of displays in which one male tries to intimidate another by charging his opponent, throwing rocks, flailing, and swaggering, and although fighting sometimes does occur, it is uncommon and seldom fatal.[1] Such encounters are usually followed by some form of reconciliation, appeasement, or reassurance in which the loser acknowledges his subordination to the winner. Franz de Waal has termed this process the "threat-submission-reassurance syndrome."[2] After an aggressive incident, both chimpanzees and rhesus monkeys have a tendency to make body contact with their opponents, usually by kissing or grooming. Such gestures are usually preceded by a beckoning hand gesture that is also used for begging for food. Among humans, handshakes are a common method of reconciliation.

Among chimpanzees, who grooms whom depends upon the social context of the encounter. In one context, the loser kisses or grooms the winner to appease him, but in another context the winner kisses or grooms the loser to assure him that the aggression has ended and that they can resume their previous relationship, albeit in the dominant-subordinate relationship that has just been established between the winner and loser.[3] The result is a form of peace agreement between the two. In a colony of pigtail monkeys (*Macaca meenestrina*), after a defeated male demonstrates submissive gestures, the winner terminates the fight, grooms the loser, and licks his wounds.[4] The establishment of a peaceful relationship between a winner and a loser is thus conditioned upon the formal acknowledgment of the winner's dominance over the loser, and submission is an effort by the loser to establish a friendly and harmonious relationship with the winner. Among rhesus monkeys, the recognition of dominance within the community is essential for the social integration of the community.

Reconciliation after a fight or nonviolent conflict between humans may involve handshakes, kisses, embraces, gentle touching of the other's arm, gift giving, inquiry about the other's feelings, making a joke, a suggestion for collaboration, or a statement of friendly intent. Children sometimes give each other emotional support by assuming fantasy roles.

In most societies, dominant individuals are usually responsible for regulating and controlling tensions within the society. Ideally, after a new dominance relationship has been established within a community, the dominant individuals become conciliatory and forgiving and assume a tolerant attitude that can create room for a degree of assertiveness, protest, and even teasing by subordinates, so that the dominance relationship becomes a way of avoiding tensions. In most human organizations, subordinates will often do things to enhance the position of their superior to make him feel more important and in turn, the superior makes gestures to show that he cares for and respects his subordinates. Such actions strengthen the bonds of loyalty between the superior and the subordinates.[5]

Reconciliation appears to be an acquired skill rather than an inherited one. In an experiment, juvenile rhesus monkeys that were relatively aggressive and intolerant with low levels of reconciliation were housed, for five months with genetically closely related stump-tailed monkeys (*M. arctoides*) juveniles—a highly conciliatory species. The rhesus monkeys were then returned to their natal group and observed for six weeks. The rhesus monkeys that had been exposed to the stump tails were three or four times more conciliatory than those that had not.[6]

The Robbers Cave Experiment[7]

The Robbers Cave Experiment provides some interesting insights into how competition among human groups arises, and how reconciliation between competitive groups can be achieved after a conflict. In the summer of 1954, a group of sociologists from the University of Oklahoma took two groups of 11 year-old boys to the Tom Hale Boy Scout Camp outside of Oklahoma City for an experiment in intergroup conflict and reconciliation. The experiment consisted of three phases, each lasting about a week. During the first phase, the two groups were kept isolated from each other at separate campsites so they could bond together into two cohesive groups. During the second phase, the two groups were allowed to interact in a competitive environment during which considerable hostility and animosity was allowed to develop between them. In the third phase of the experiment, the two groups were presented with a series of tasks, the successful completion of which required the cooperation of both groups to complete.

The participants were 22 well-adjusted white Protestant fifth grade boys from different public schools in Oklahoma City who did not know each other prior to the experiment. They were all of average IQ, physical development and athletic ability. The two groups were matched as closely as possible in terms of height, weight, general athletic ability, sports skills, popularity at their schools, musical and skit skills, swimming ability and previous camping experience. The two groups were transported in separate busses to the Scout Camp and housed in separate cabins about a half mile apart. They shared a common mess hall but ate at different times during the first phase of the experiment. Because the area was heavily wooded, neither cabin was visible from the other, nor from the mess hall, which was located approximately halfway between the two cabins. Each group had its own bathhouse, swimming and campfire facilities, and a canoe.

During the first week, the two groups lived and played together in their respective camps and bonded into two cohesive groups. Each group adopted a name: Rattlers and Eagles. Each group had a recognized leader and had accepted its leader's suggestions for the baseball team's captain, and suggestions by other members of the group were adopted only after the approval of the group's leader. Both groups had established hierarchies that remained relatively stable throughout the experiment, and each group became territorial toward its cabin site and swimming facility. When the Rattlers heard the Eagles playing at the common baseball diamond, they considered it to be an intrusion onto "our ball field," and one boy said, "They better not be in our swimming hole."

Over the course of the first week, each group became aware that the other group was present in the camp when they heard the other playing baseball, and each group was eager to challenge the other to a baseball game. Even before they met for the first game, the Rattlers had developed a highly competitive attitude toward "those boys at the other end of the camp." At the beginning of the second phase of the experiment, each group was told that a tournament would be arranged between it and the other group with a trophy, medals, and jackknives as prizes. The tournament consisted of ten events, which included baseball and touch football games, tug-of-wars, a tent-pitching contest, skits and songs, a treasure hunt, and cabin inspections.

The first event of the tournament was a baseball game. The Rattlers arrived at the ball diamond first, which they considered to be theirs, and they considered themselves the home team. After the Eagles arrived, the two groups looked each other over. One Eagle made a derogatory comment, and he was answered with a catcall from the Rattlers, and razzing began. As the game began, the Rattlers began singing, "The first Eagle hit the deck, *parler vous* .. . The second Eagle hit the deck, *parler vous* . . ." and the Eagles responded, "Our pitcher is better than yours. . . . Our catcher is better than yours." As the game progressed, the Rattlers referred to one of the Eagles as "Fatty" or "Tubby." The Rattlers responded to the razzing by calling, "You're not eagles; you're pigeons." After the Rattlers won the game, the Eagles put on a display of good sportsmanship. The two groups ate at the same time following the game, but at separate tables in the mess hall. There was considerable razzing and food throwing between the two groups, and they alternately sang derogatory songs. Some of the Eagles expressed their desire not to eat with the Rattlers again.

The next event was a tug-of-war, held after supper on the second day of the tournament. After the Eagles pulled the first Rattler over the line, the Rattlers started pulling the Eagles and continued until they had pulled all of the Eagles over the line. When one of the Eagles saw that his team was losing, he walked away from the rope. The winning Rattlers cheered, jumping up and down and slapping each other on the back, they then gave three cheers for the Eagles, and then walked past the dejected Eagles with much yelling and razzing. The Eagles stood around talking about "how big the Rattlers were."

As the Eagles were walking back to their cabin, they noticed that the Rattlers' flag was still on the ball field's backstop, and they yelled that they should take it down. Craig (E) * climbed the backstop and tore it down. Mason (E) grabbed it and tried to tear it up but couldn't. Someone then suggested that they burn it, and they set the Rattlers' flag on fire and hung the scorched remnants of the flag on the backstop. Mason said, "You can tell those guys I did it. If they say anything, I'll fight 'em."

The next morning, the Rattlers ate their breakfast first and were the first to arrive at the ball field. When they saw the burnt remnants of their flag, their reaction was noisy and resentful, and several suggestions were made for retaliation against the Eagles, so they devised a plan: When the Eagles arrived at the ball field, Simpson (R) would ask if they had burnt the flag. If they admitted that they had, Simpson would start a fight and the rest of Rattlers would join in. When the Eagles arrived at the ball field, they admitted that they had burned the flag, and the Rattlers yelled invectives at the Eagles. Martin (R) grabbed the Eagles' flag and ran down the road with it, with Mason (E) in pursuit. The Eagles then grabbed the Rattlers' second flag and tore it up. When the Rattlers tried to retrieve it, Swift (R) grabbed Craig (E) and held him in a wrestling hold, asking which Eagle had burned the flag. Craig (E) replied that they all had. After Simpson (R) started a fist fight with Cutler (E), a staff member stopped the fights. Mason (E) yelled for someone "my size" to whip. Mills, the Rattlers' leader said, "Here I am," but the staff stopped the fighting.

The staff then started the second ball game over the Rattlers' violent objections to the Eagles being named the "home team" that day, yelling, "We built everything but the backstop." The game finally got underway with considerable razzing and name calling. The Eagles won

---

* (E) designates an Eagle, (R) designates a Rattler replied, "Here I am.", and

the game and afterward, two of the Eagles said, "Nice game" to the Rattlers. The next contest was another tug-of-war which, this time, the Eagles won. The Rattlers then planned a raid on the Eagles' cabin that night. The raid took place at 10:30 that evening after the Eagles had gone to bed. Some of the Rattlers entered the Eagles' cabin, turned their beds over, and ripped the screens. The next morning, the Eagles retaliated while the Rattlers were at breakfast. The Rattlers then won a touch football game and the first tent pitching contest. The last event of the second phase was a raid by the Eagles on the Rattlers' cabin.

By the end of the second phase of the experiment, neither group wanted anything to do with the other. Invectives and names had been hurled back and forth—"stinkers," "braggers," "sissies"—and both groups objected to eating with the other. The members of each group had formed negative attitudes toward every member of the other group. At this point, the animosity between the two groups had reach a peak, and the in-group cohesion of both groups had also peaked.

The objective of the third phase of the experiment was to eliminate the hostility between the two groups—that had become very intense during the second phase—and to achieve a reconciliation between them. During the first two days of the third week of the experiment, the two groups of boys were given seven opportunities to mingle together during meals and shooting firecrackers on the Fourth of July, but during these episodes, the two groups stayed apart and there was no contact between their members. One of the experimenters observed at the end of the second day, "It was apparent by this time that mere contact between the groups without the interdiction of joint goals was going to be insufficient to reduce the negative relationship between the groups."

The camp's water was supplied from a reservoir on a mountain north of the camp and was pumped into a large tank located on another mountain about 1¼ miles from the camp. There was a valve at one end of the tank that controlled the flow of water from the tank to the camp and a faucet at the other end. Before this event, the experimenters turned off the valve, placed two large stones over it, and stuffed rags into the faucet. The boys were then told that there was a possibility of a water shortage at the camp and were advised to fill their canteens with water that remained in the system.

The boys were then assembled into one group and told that the system had completely failed and that the difficulty might be a leak in the pipe between the tank and the camp or that it might be at the pump, the reservoir or the tank. The area between the camp and the reservoir was to be inspected by four crews. As volunteers for the four crews were called for, boys from the same groups stepped forward, so that each crew contained only boys of the same group.

If no problems were found, all the crews were to assemble at the tank. In a little over an hour, having found nothing, all four crews had congregated at the tank. All the water in the canteens had been drunk, and when the faucet was turned on, nothing came out. A ladder was found near the tank, and the boys took turns climbing to the top of the tank and found that the tank was three-quarters full. Attention then returned to the faucet, and one of the boys found the rags in the faucet. Using their jackknives, the Eagles and Rattlers alternated in trying to remove the rags. Finally, one of the Rattlers suggested that the Eagle staff observer was a "big guy" and that he should try. The suggestion was seconded by all the boys, and water finally gushed from the faucet. When the enthusiasm of clearing the faucet had died down, the boys started drifting away from the faucet, and there was a noticeable increase in mingling between the two groups that took place within the framework of a common deprivation that affected everybody. A group of Eagles then went to the other end of the tank, removed the boulders, and turned on the valve that supplied water to the camp.

The animosity between the groups returned at supper that evening, but the next morning, one of the Rattlers said, "Howdy," to the Eagles, and several of the Eagles returned the salutation, and they began to sing the Caisson song together. As the boys were putting food on their trays, the cook called for their attention and said that cleaning up after the food and bottle-cap throwing was taking so much time that she wouldn't be able to cook such good meals for them if they didn't stop the throwing. The appeal was effective. During breakfast, someone threw an object, and one of the Eagles said, "Remember, you guys, no throwing," and the throwing stopped.

Up to this point, the Eagles had always entered the mess hall first, with the Rattlers calling, "Ladies first," but as they were waiting for breakfast the second morning after the incident at the water tank, the groups reached an agreement that the Rattlers would enter first for breakfast and the Eagles first for lunch. Thus, the notion of taking turns was introduced at the intergroup level, to regulate matters of mutual concern rather than each group rushing to be first.

The next event was a campout to a lake about 14 miles south of the Scout camp. When they arrived at the lake, the boys went swimming, after which they were hungry for lunch. The food for their lunch had been left on another truck and had to be retrieved. As the boys watched in two groups about 15 yards apart, the truck driver pretended that he couldn't start his truck. One of the boys suggested that they push the truck to get it started, but the truck was pointing uphill. Another boy spotted their tug-of-war rope on the truck and shouted, "Lets have a tug-of-war against the truck. Twenty of us can pull it for sure." They threaded the rope through the truck's bumper so each group could pull on an end. By the second attempt to start the truck, the two groups had become thoroughly intermixed, and finally, the truck started. They all shouted, "We won the tug-of-war against the truck." There was intermingling and backslapping. Four boys, two from each group, then went to the water pump and pumped water for each other.

After the food arrived, it was placed on a table between the two campsites, and the Rattlers and Eagles gathered around the table and began discussing whether the two groups would alternate in food preparation or each group would prepare its own meals, but in the midst of the discussion, they unconsciously began to work together at preparing their lunch. Then they lined up to serve themselves, the first two Eagles through the line sat at a centrally located table. The first three Rattlers sat at a table located at their camp site, but the next Rattler went to the central table and sat with the two Eagles. After finishing their meals, two of the Rattlers drifted over to the table where the Eagles were sitting.

That afternoon, the boys went swimming. The Rattlers went first, but when they saw a water moccasin (a snake) darting about in the water, they got out of the water. When the Eagles arrived at the pier, the two groups got into an excited discussion about the snake and then went to another location and swam together.

The next task was to pitch their tents, but the components of the tents—poles, stakes, hammers—had been dumped in two piles and had not been sorted, so that when the boys started to pitch their tents, they didn't have some of the needed parts and had an excess of others. Both groups took their extra parts to a central location and matter-of-factly started to exchange the parts. That evening, the boys casually intermingled to jointly prepare their supper.

A trip to Arkansas had been planned for the following day, but on the morning of the planned trip, the boys were told that the Eagles' truck, which the boys had started by towing it in the "tug-of-war," could not make the trip, and the decision of what to do was left to the boys. They could make two trips in the Rattlers' truck, or they could abandon the trip to Arkansas and go back to the Scout camp. Finally, one of the Rattlers suggested that the Rattlers' truck be unloaded so that all the boys could travel together. After some discussion, they agreed to go together in the Rattlers' truck. During the trip, the boys spontaneously began alternating singing songs that had been associated with the two groups. At rest stops and during lunch, there was thorough intermixing of the two groups, with very few signs of group demarcation. During one rest stop, agreement was reached that the boys could return together to Oklahoma City the next day in one bus, rather than in two as had been originally planned.

Before a planned refreshment stop on the trip back to Oklahoma City, a Rattler proposed that the five dollars that they had won in the tournament be used to buy milk shakes for all the boys. The rest of the Rattlers enthusiastically agreed. During the stop, one of the counselors reminded the Rattlers that the five dollars would pay for all of the Rattlers' refreshments if they didn't buy milk shakes for the Eagles, but the Rattlers replied that they were glad that they had decided to share their prize. During the remainder of the trip, general gaiety prevailed as the boys sang "Oklahoma," exchanged addresses and telephone numbers, and agreed to meet again. At the end of the experiment, the two groups that had developed deep hostilities and animosities toward each other had reconciled their hostilities and had merged into a single coherent group, probably because they had cooperated in superordinate tasks to achieve goals that all the boys desired but that could not have been achieved by either group alone.

War and Reconciliation in the Twentieth Century

World War I Peace Settlement. The settlements that followed the First and Second World Wars offer a contrast between punishment and reconciliation. When Germany signed the Armistice Agreement on November 11, 1918, it believed that the expected peace treaty would be based on President Woodrow Wilson's fourteen points and his subsequent pronouncements. In the pre-armistice exchange of notes between Germany and the Allies, Germany had implied that the object of the ensuing peace conference "would be to agree upon the practical details of their (Wilson's fourteen points) application."[8] Although Wilson pressed for a just peace based on his fourteen points, France and England had other ideas. Both wanted Germany to pay reparations for their war costs, and France, fearful of a future attack by Germany, also wanted Germany to return the territories west of the Rhine River, including the Rhineland, which Germany had taken from France during the 1870–1871 Franco-Prussian War, Alsace-Lorraine, and the coal-rich Saar region.

The Peace Conference met in the Salon de Paix at the Quai d'Orsay in Paris between January and May 1919. Initially, it included all twenty-seven of the Allied Powers, but excluded all of the Central Powers and Russia, that had not been invited to the peace conference. But after the middle of March, the conference was limited to the Big Four—Wilson (the United States), Clemenceau (France), Lloyd George (England), and Orlando (Italy)—which, by the end of April, had agreed among themselves on the terms that Germany would be forced to accept. The alternative would be the resumption of the war. The terms of the Versailles Treaty included: Alsace-Lorraine would be returned to France; France would recover the mines in the Saar that Germany had flooded during the closing days of the war, and the Allies would occupy the Saar and the Rhineland for the next 15 years. Germany would not be allowed to build fortifications on the west bank of the Rhine or within 50 kilometers of the east bank. In the east; Germany would cede about 75,000 square kilometers (30,000 square miles) to a recreated Poland.

The treaty's military terms included: the German Army would be limited to 100,000 men, hardly the size of a police force for a country the size of Germany; army officers would be required to serve 25 years to prevent the creation of a large reserve officers corps; and the German General Staff would be dissolved. The navy would be limited to six battleships of no more than 10,000 tons, six light cruisers, twelve destroyers, and twelve torpedo boats. The remainder of Germany's navy would be scuttled at Scapa Flow in the Orkney Islands in Northern Scotland. Germany could possess no submarines, military aircraft, tanks, heavy artillery, or poison gas.[9] Under the financial terms of the treaty, Germany would pay an unspecified amount of reparations to be determined by a special Reparations Commission, which were set at $54 billion in 1921. Earlier, the British had asked for $120 billion, and the French for $ 220 billion, as compensation for their war losses.[10] The peace terms also included a "War Guilt" clause, that stated, "The Allied and Associated Governments affirm and Germany accepts the responsibility of Germany and her allies for causing all the loss and damage to which the Allies and Associated Governments and their nationals have been subjected as a consequence of the war imposed upon them by the aggression of Germany and her allies."

These terms were written into the Versailles Treaty, which was given to the German delegation on May 7 that was told that it had two weeks to respond with written comments; there would be no oral discussion of the treaty. After the German comments were submitted, a few minor adjustments were made to the treaty, but the terms remained substantially unchanged. The German "peace settlement" was not a reconciliation; it was a punitive imposition to punish Germany not only for its own crimes, but also the crimes of others, including those of the Allied Powers "This is not a peace; it is an armistice for twenty years," remarked Marshal Ferdinand Foch, the supreme allied commander, and Matthias Erzberger, the head of the German Armistice Commission, said, "Do not expect us to be our own executioners."[11]

The Interwar Period, 1919–1939. The German people did not accept their defeat or war guilt. They believed that Germany's collapse in November 1918 was due to the Allied naval blockade that had deprived Germany of its means to fight, Allied propaganda, and political disunity within Germany. They even denied that Germany had suffered a military defeat, or that Germany had any responsibility for the outbreak of the war. Within Germany neither the conservatives nor the army acknowledged the Versailles Treaty, and the army was able to regain the power and status that it had had within Germany before the war. The army became a state within a state. Germany never fully paid her reparations debt. These attitudes were largely responsible for Adolf Hitler's rise to power and the outbreak of the Second World War.

Adolf Hitler had been a corporal in the German army during the war. He was wounded twice and was twice awarded the Iron Cross. Hitler was a rabble-rouser, a fanatic, but a compelling public speaker, and he also had a knack for organization. He grasped the strength of the new nationalist, antidemocratic tide that was rising in Germany after the war. In November 1919, Hitler joined the German Workers' Party in Munich, an ultra right wing group with scarcely a hundred members that espoused a strong nationalist Germany and the overthrow of the democratic Weimar Republic. Because of his oratorical skills, Hitler soon became the party's leader, and he renamed the party the National Socialist German Workers' Party, that became known as the Nazi Party. He then organized a quasi-military adjunct to the party—the Storm Troopers or Brown Shirts—whose official name was the Gymnastics and Sports Division of the Party to avoid suppression by the federal government.

In November 1923 Hitler led the "Beer Hall Putsch" in Munich in which his Brown Shirts surrounded a beer garden. Hitler then entered the hall and jumped on to a table, fired a pistol into the air and declared, "The national revolution has begun," and announced that the Bavrian and federal governments had been removed and that he had formed a provisional government for Germany. He was arrested, tried for treason, and sentenced to five years in the Landsberg fortress. But he spent only nine months there during which he wrote the fist volume of his notorious *Mein Kanpf* (My Struggle). In it he described a future Germany that would become the "Lord of the Earth." It would be a dictatorship, and he declared that Germany must expand, mostly to the east at the expense of Poland and Russia, but first it must take all of the lands that had been taken from Germany under the terms of the Versailles Treaty. France was the "inexorable, mortal enemy of the German people." The aims of the French toward Germany, he declared, were a "dismembered and shattered Germany, a hodgepodge of little state s." There must be "a final active reck oning with France . . . a last decisive struggle," he declared. He envisioned a Germany that would occupy all of Europe and most of Russia. He declared that the Aryan race was the master race and that it would become supreme by trampling others. Slavs and Jews were chaff. Czwchs, Poles andand Russians would become "hewers of wood and drawers of water for their German masters."[12]

The membership of the Nazi Party had risen to 27,000 by 1925, and to 178,000 in 1929. After 1929, Hitler's rise was fueled by the humiliation of the Versailles Treaty, the inflation that Germany had suffered during the 1920s, and the worldwide depression triggered by the collapse of the US stock market in October 1929. Millions of Germans were out of work, as were millions of others in other western countries. In the May 1928 elections, the Nazis polled only 810,000 votes and elected only 12 members to the Reichstag, but in September 1930, the party polled 6.4 million votes and took 27 seats in the Reichstag. In 1932, Hitler ran for president of Germany against the popular but senile Paul von Hindenburg and polled 13 million votes to Hindenburg's 19 million. Then, as the Depression worsened, the government of General Kurt von Schleicher fell, and Hitler persuaded Hindenburg to appoint him chancellor.

Hitler next set out to seek revenge for Germany's 1918 defeat and humiliation under the Versailles Treaty, and in the process, he set the world on fire. In October 1934, he increased the strength of the German Army from its authorized strength of 100,000 men to 300,000, withdrew Germany from the League of Nations, authorized the construction of two battle cruisers of 26,000 tons, and initiated the construction of submarines in Finland, Holland, and Spain. France and England stood by and did nothing to stop him. In February 1935, the population of the Saar voted overwhelmingly to rejoin Germany, and Hitler proclaimed that he had no further claims on France, but in March 1936, German troops occupied the Rhineland and reclaimed it for Germany. Two years later, Hitler marched his troops into Austria and annexed that country to the German Reich. The Austrian people welcomes the German troops with Nazi salutes, Nazi flags and flowers.

The story of the Nazi occupation of Czechoslovakia is a long and painful one. In June 1938, Hitler set October 1 as the day that Germany would occupy the country, but on September 30, Europe's leaders met in Munich and agreed that Hitler could occupy 28,000 square kilometers of the Sudetenland, Czechoslovakia's western province that contained 2.8 million ethnic Germans. When the English prime minister, Neville Chamberlin, returned to London, he declared that he had achieved "peace in our time." The day following the Munich conference, German troops occupied the Czech Sudetenland. Then Poland and Hungary swept in like vultures: Poland occupied some 1,500 square kilometers in the north of Czechoslovakia, and Hungary some 19,000 square kilometers in the south. Finally, on March 15, 1939, Hitler occupied the rest of the country.

World War II. The final blow to Europe's peace was struck on September 1, 1939, when Germany and Russia simultaneously invaded Poland, igniting the conflagration of the Second World War. Germany and Russia had arranged to divide Poland between them in the secret Molotov-Ribbentrop Pact signed on August 21, 1938, but in September 1941, Germany invaded Russia. Before that, Germany had invaded France and the Low Lands in May 1940, occupied the northern half of France, and established the Vichy puppet regime in southern France. Italy invaded and occupied the southeastern portion of France.

The tide of the war turned in 1944 when on June 6, the Americans and British landed an amphibious force of 300,000 men on the Normandy coast and dropped another 24,000 airborne troops on France. In the east, the Russian Army had broken the 900 day siege of Leningrad (now St. Petersburg) in January 1944 and liberated the Crimea and Ukraine in May. The Russian Army drove the German Army back to Berlin and then occupied all of Eastern Europe. The war ended in April 1945 when the Allied Forces converged on Berlin. Hitler and his close associates committed suicide in their bunker in the Bavarian Alps.

The European Recovery and Integration. The Second World War was the most devastating war in European history. More than 50 million people lost their lives during the war, and, unlike in the First World War, most of Europe's industry had been destroyed by the strategic bombing campaigns of both the Allies and Germany. Furthermore, in 1945, the Soviet Army occupied all of Eastern Europe, including the eastern halves of Germany and Austria, and all of Poland, Czechoslovakia, Hungary, Rumania, Bulgaria and the Baltic States of Estonia, Latvia and Lithuania. Industrial production was at a standstill because of the lack of skilled labor, agricultural production was only half of what it had been before the war, and food had to be rationed. Because bridges, railroads, and harbors had been prime targets of the bombing campaigns, most had been destroyed, and transportation in all of Europe was at a standstill. The winter of 1946–1947 was a particularly harsh one for Europe. In England, the shortage of coal was so severe that the government had to reduce its delivery to industry to 50 percent of its prewar levels, and people wore their overcoats in their homes to keep from freezing. Not only not was enough coal being mined, but there was a shortage of railway wagons to deliver what was mined.

Because of the harsh winter, an estimated 3.5 million acres of wheat planted in the autumn was destroyed in January and February, confronting Europe with the prospect of importing two to three million tons at $100 a ton. Farmers were not sending their produce to markets, and the cities were slowly starving. A *New York Times* survey of the food situation in Europe concluded that famine and severe food shortages confronted every European country. The communists exploited these conditions in France and Italy, where their voting strength was about one-third of the total and where they occupied prominent cabinet posts.

It was clear in both Europe and the United States that it would be impossible for the Europeans to recover without massive help from the United States, the only country that had not suffered the devastation of the war. American farms and factories were running at full force. The United States had already provided $10 billion in aid and loans to Europe from 1945 to 1947, but Secretary of State George Marshall recognized that much more was needed. He asked his deputy, Dean Acheson, and his counselor, George Kennan, to devise a program for the recovery of Europe, and he announced it in his Harvard commencement speech in June 1947. He proposed that the countries of Europe should collectively devise a recovery plan, and he committed the United States to loan Europe the funds needed to implement the plan. The plan would be open to all of the countries of Europe, including the Soviet Union and its occupied satellites. The original State Department proposal envisaged the creation of a united Western European Federation.

Both the British and French grabbed at Marshall's offer and approached the Soviets and the occupied countries of Eastern Europe to participate, but Stalin forbade his satellites to attend the organizing conference in Paris. In July 1947, 16 European countries met in Paris and established the Committee on European Economic Cooperation. They listed their separate needs and asked the United States for $29 billion, much more than the United States had anticipated. The European Recovery Plan—the Marshall Plan—provided Europe with a little more than $12.5 billion to implement the plan. In April 1948, the same 16 countries formed the Organization for European Economic Cooperation (OEEC) to coordinate the recovery program. The objectives of the OEEC included the elimination of the restrictions on trade, the reduction of tariffs among the states, and the creation of a free-trade area or customs union among its members.[13]

To be successful, the Marshall Plan required the participation of Germany—Europe's old nemesis—because it was the country with the greatest economic and industrial potential. The initial policies of the Western Allies toward Germany were denazification, disarmament, and democratization—the three D's. After the war, Germany and Berlin, which was occupied by Soviet Russia, were each divided into four occupation zones, administered by the United States, the United Kingdom, France, and the Soviet Union. The German people also suffered from shame of starting the war and because of the atrocities they committed during the war,[14] but instead of punishing their former enemy, as they had in 1919, Germany's former enemies in the West offered Germany's western zones reconciliation and integration into the European community. As the first step toward rebuilding Germany's self-sufficiency, her former enemies in the West agreed in June 1948 to create a new West German state from their three occupation zones.

The Soviets responded with the Berlin blockade and cut all road and rail communications between West Berlin, which lay in the Soviet occupation zone, and the western zones. To prevent the West Berliners from starving, the Western powers initiated the Berlin Airlift, which delivered 3,000–4,000 tons of food, fuel, and other necessities every day to the two million people of West Berlin between January 1948 and June 1949, when the Soviets lifted their blockade. I

n response to the Berlin crisis, the Western Powers created the North Atlantic Treaty Organization (NATO), in which the western Europeans, together with the United States, Canada, Turkey and Iceland, agreed to form a defensive alliance to counter the threat posed by the Soviet troops that were stationed in Eastern Europe. But to provide an effective defense against the Soviet troops, western ground forces would have to be stationed in the newly created West German state, and thus Germany would have to be an integral member of NATO. Italy, Germany's ally during the war, was also welcomed into NATO and the European community.

In May 1951, the French foreign minister, Robert Schuman, proposed that the production of coal and steel in France and Germany be placed under a common authority, saying that it would be the "first step in the federation of Europe" and would make war between France and Germany "not merely unthinkable, but materially impossible." Six nations signed the Treaty of Paris that created the European Coal and Steel Community (ECSC), which is now widely considered to have been the first step toward the integration of Europe into a single political and economic entity. Then in March 1957, the six members of the ECSC signed the Treaty of Rome, which created the European Economic Community, commonly known as the Common Market, which was another step toward the creation of a European political union. The members of the Common Market agreed to harmonize their economic policies and to eliminate all restrictions on the movement of peoples, services, and capital within 12 years.

In the fall of 1991, large numbers of East Germans began leaving East Germany through Austria while other East Germans dismantled the

Berlin Wall. The US president, George H.W. Bush, with Mikhail Gorbachev, the president of the Soviet Union, effectively managed the dissolution of the Soviet Union and the Warsaw Pact into 20 newly independent states.

The final step in the integration of Europe was taken in 1992 with the signing of the Treaty of Maastricht, which created the European Union (EU). It created a common European currency—the euro—and a common defense policy, and established a common citizenship for all the citizens of the EU. By January 2012, the EU had grown from its original 12 states that signed the Treaty of Maastricht to 27 states, including eight former members of the Soviet Bloc. The EU has a common currency that is used by only 17 of its member states, but each state continues to maintain its own fiscal and political policies.

The history of Europe in the twentieth century illustrates how a deep hostility harbored by Germany towards its neighbors that was the result of the Versailles Treaty and resulted in the tragedy of World War II can become the cooperative effort that eventually lead to the creation of the European Union. After the end of World War II it became apparent to the victorious allies that the recovery of Europe would not be possible without Germany's participation since Germany had the greatest industrial capacity in Europe. Consequently the victors eagerly sought Germany's participation in the Marshall Plan, the European Coal and Steel Community and the other institutions that followed, including the formation of the West German state from the three occupation zones of France, the United Kingdom and the United States.

When NATO was established in response to the Berlin Blockade, it was clear that Western Europe could not be defended against a Soviet invasion without the participation of West Germany since Germany was the frontier state through which the Soviet troops would pass. Consequently Germany was included as a founding member of the NATO alliance.

The European Union's Stress Test. As the 21st century unfolded, the EU encountered a series of disquieting events that would severely test its viability. To consolidate the EU and to improve its governance, the EU began drafting a European Constitutional Treaty in December 2001 that was completed and signed by 25 of its member states in October 2004. After 18 of its member states had ratified the treaty, the French and Dutch citizens rejected it in their referendums. The remaining seven member states then cancelled their ratification procedures and the treaty died.

The Maastricht Treaty had created the EU with a common currency, a common defense policy, and a common citizenship but had failed to create a common fiscal policy for the EU. This meant that each of the member states would have its own welfare, labor, fiscal, and financial policies. It also meant that because of the common currency, there was no possibility of adjusting exchange rates to accommodate these differences, and Germany, which pursues less generous social policies than the rest of Europe does, would have to pay for the excesses of Greece, Italy, and Ireland. The origin of this conundrum was the concerns of the French president François Mitterrand, who was alarmed at the increased influence that the German unification had given that country and wanted to prevent Helmut Kohl, the German chancellor, from implementing a common fiscal and political policy for the EU, but Kohl maintained that "political union [with a common fiscal policy] is an essential part to economic and monetary union." But, Mitterrand objected, France wanted to gain some control over Germany's currency but did not want Germany to have any control over France's budget.[15] The solution was the "convergence criteria," which were incorporated into the 1992 Maastricht Treaty and required each member state to keep its public debt to under 60 percent of its gross domestic product (GDP), and its budgetary deficits to less than 3 percent of its GDP. The treaty contained no enforcement mechanisms. The discussion over a fiscal union withered away and became moot. In 2003–2004, both Germany and France exceeded the 3 percent limit on budget deficits, and the Mediterranean states followed with even greater deficits.

When the world financial crisis arrived in 2008, triggered by the subprime mortgage crisis and the fall of the Lehman Brothers Bank in the United States, public spending in Greece and Italy skyrocketed, whereas in Spain and Ireland, private spending rose for vacation homes, BMWs, and Bosch washing machines because both governments and individuals could borrow at unprecedented low rates. Staggering under the financial burden of its unification, Germany cut its labor costs and trimmed its welfare costs, while France and the Mediterranean states allowed their labor costs to soar by cutting working hours and lowering retirement ages.

The Germans began to complain, "Why should we work even harder and retire even later so these feckless Greeks, Portuguese, and Italians can retire earlier than we do and go sun themselves on the beach?" The German tabloid *das Bild* advised, "Sell your islands, you bankrupt Greeks." As the investors began to realize the isks associated with the bonds of the peripheral states, interest rates on their bonds rose to 5 and 8 percent, putting increasing strain on Greece, Portugal, Italy, and Spain.[16] Greece and Italy were forced to appeal to the European Central Bank and the International Monetary Fund (IMF) to bail them out, but the IMF imposed such harsh terms on both countries as to cause riots that forced governments to fall and new elections to occur.

The crisis struck Greece the hardest. Its public debt had risen to 180 percent of its GDP, almost twice the average for the rest of Europe and tax evason had become a national pastime in Greece. In 2014 the government had collected less an half of the taxes due. As the crisis grew, Greek's creditors were forced to extend their loans to Greece, but at a cost of more and more austerity measures on Greece which its population resisted. In January 2015 Greece elected Alexis Tsiprus as prime minister who has continually defied Greek's creditors including the European Central Bank. Greek's creditors had no choice but to continue extending the Greek debt. In the meantime Greece suffered high inflation and the inability to purchce foreign goods or to sell its goods and serices abroad.

Because of the crises, any spirit of a unified Europe that had ever existed disappeared, and pressures toward unification greatly receded. Turkey and Serbia, which once had been clamoring at Europe's door, withdrew their applications and no longer wanted to join the EU. Europe is still a collection of 27 diverse cultures and 26 languages, and a spirit of Europeanism has ceased to exist except among a few bureaucrats and elites.

Failed Peace Negotiations

Not all efforts at reconciliation have been successful. We have already seen how the peace negotiations after World War I, in which the Allies imposed a punitive peace on Germany and its allies resulted in the rise of Hitler and the outbreak of the Second World War, but in at least three other cases—the Good Friday Agreements to settle the conflicts in Northern Ireland, the Israeli-Palestinian conflict and the Dayton Accords to resolve the conflicts between the ethnic groups in Bosnia—attempts to resolve the conflicts either were only partially successful or failed.

Northern Ireland. After more than four decades of violent conflict between the Protestant majority and the Catholic minority in Northern

Ireland in which more than 3,600 people had been killed, the Irish Republican Army (IRA), the military wing of the Sinn Fein Catholic party, announced in August 1994 that it would cease its military operations because Sinn Fein had concluded that the British government had little interest in continuing its administration of Northern Ireland and that Sinn Fein could achieve its aims through nonviolent political action. In 1996, George Mitchell, a former US Senator, was asked by President Bill Clinton to mediate between the contending parties in Northern Ireland and the British and Irish governments. Senator Mitchell's efforts resulted in the 1998 Good Friday Agreements, which established procedures for forming a power- sharing government in Northern Ireland. A joint government was established in June 1998, but it was dissolved in 2002 because, first, refused tom decommission its arsenal of weapons; second, there were allegations of IRA espionage activities; and third the violence in Northern Ireland had continued, although at a lower level than before.

In July 2005, Sinn Fein and the IRA announced for the second time that they had given up their armed struggle and would enter into a political process to reestablish the power sharing government. In October of 2006, the Decommissioning Commission concluded that the IRA had fully renounced terrorism and had committed itself to following a political path. The British and Irish Prime Ministers, Tony Blair and Bertie Ahern, called the feuding Catholic and Protestant leaders in Northern Ireland together in October 2006 and offered a compromise formula—the Saint Andrews Agreement—to restore the power sharing government. They told the Northern Irish parties that they had until November 24, 2006, to agree to join in a power sharing government. When Gerry Adams, the Sinn Fein leader, and Ian Paisley, the Protestant leader, hesitated, Peter Hain, the British Secretary for Northern Ireland, who was also acting as the governor-general for Northern Ireland in the absence of a functioning government in Belfast, sought to make direct rule from London as unpalatable as he could and to achieve an agreement he decided to use "tough love." He threatened to introduce controversial water bills and unpopular reforms of Northern Ireland's school system, and to cut off the salaries and allowances of the politicians if a power-sharing agreement was not reached by March 26, 2007. Finally unless an agreement was reached by the deadline, the devolution of power from London to Belfast would be suspended indefinitely. Because these terns were totally unceceptable to both the Protestants and Catholics, Paisley and Adams agreed to accept the terms of the Saint Andrews Agreement and to form a coalition government.[17] The threat to impose direct rule from London on Northern Ireland was the primary factor in bringing the two parties together in a coalition government.

While a formal peace remained in place in Northern Ireland, Catholic and Protestant neighborhoods in Belfast were separated by "peace walls" that kept Catholics and Protestants segregated. Occasional violence continued at a low level between Protestant and Catholic youth groups, and Protestant Orangemen continued to march through Catholic towns and neighborhoods during the summer marching season.

This uneasy truce continued until December 2012, when the pro-British Unionist members of the Belfast Council were outvoted by a coalition of Catholic Republicans and a small liberal bloc organized to limit the flying of the British Union Jack on the Council building to 18 days a year. The 18 days did not commemorate any historical events such as Nelson's victory at Trafalgar in 1805, or Germany's surrender in 1945, but were the birthdays of the Queen and her family members. Since 1906, when the Council was dominated by Protestant Unionists, the flag had flown from the top of the building every day of the year. Under Britain's strict rules, the English flag was flown on both public and private buildings, including the Parliament building, only on the designated days, but applying the rule in Belfast provoked a violent reaction by the Protestant Unionists.[18]

The pro-British Unionists saw the episode as part of the erosion of the British presence in Northern Ireland and as stripping Northern Ireland of its British identity. On Belfast's Shankill Road in a run- down Unionist stronghold that was notorious as a center of ambushes and bombings during the Time of Troubles, the Union Jack was everywhere: atop buildings, in shop windows, in bars and cafes, and in the hands of small children out shopping with their parents. The humiliation of the flag issue was the center of conversation everywhere. "It's our flag, our identity. It's been flown over City Hall every day since 1906, and it's being stripped from us," said Peter Shaw, owner of the Shankill Band Shop, which sold thousands of flags and other Unionist memorabilia.[19]

Underlying the dispute was the demographic in the shift in the population f Northern Areland. Between the 2001 and the 2011 consuses, those who listed themselves as Protestant or being brought up as Protestant decreased from 53 percent to 48 percent, whereas those who listed themselves as Catholic or being brought up as Catholics increased from 44 percent to 45 percent. The Good Friday Agreement specified that Northern Ireland would remain part of Great Britain as long as a majority of populations of Northern Ireland and the Irish Republic wished it to be.[20] Patricia McBride, a Catholic whose father was killed by a Unionist gunman, and was a leader in the reconciliation effort under the Good Friday Agreement, said that she feared that the peace process might founder when the Protestants realized that the population numbers were moving against them: "There was always going to be something tht triggered the upheaval. Increasingly they felt abandoned by the state whose agents they had been for so long."[24]

The Failure of Peace in Bosnia-Herzegovina. Bosnia-Herzegovina had been a republic within the former Yugoslavia. Its population consisted of three major ethnic groups: Muslims, also called Bosniaks, (44 percent) occupied the western part of the country, Orthodox Christians (31 percent) occupied the eastern part of the country; and, Roman Catholic Croats (17 percent) occupied the north. The three ethnic groups had lived and worked together for centuries and have intermingled with considerable intermarriage among them.

In October 1991, Bosnia-Herzegovina's parliament passed a Memorandum on the Sovereignty of Bosnia-Herzegovina declaring its independence by a simple majority, but its Serb members objected and the Arbitration Commission of the Peace Conference on Yugoslavia ruled that a referendum was required for independence. The independence referendum was held in March 1992, in which 92 percent voted in favor of independence, but the turnout was only 64 percent because the Serbs boycotted the referendum. The Serbs then defected and formed their own Republika Srpska, which almost completely surrounds the Croat-Muslim areas of the Bosnian Federation as surrounds the Croat-Muslim areas of the Bosnian Federation as shown in Figure 11.1

Republika Srpsk's army was commanded by General Ratko Mladic who began a brutal campaign of ethnic cleaning of the Bosniaks. Mladic's army included elements from the Yugoslav People's Army and Serbian police units. Mladic would attack towns and villages

Figure 11.1 – Bosnia-Herzegovina[21]

in eastern Bosnia and after the towns were in their hands, they began a brutal campaign of atrocities, including massacres and mass rapes. They would ransack and burn down homes and apartments, kill the men and boys and rape the girls and women. In Foca, they left 22,500 refugees and in Visegrad, they murdered 3,000 Bosniaks. Mladic established a siege around Sarajevo in April 1992, and using heavy weapons including, mortars, tanks, artillery and heavy machine guns, killed 13,950 of Sarajevo's population, including 5,434 civilians. The siege was the longest since the siege of Leningrad during World War II. It lasted 1,425 days until February 29, 1996. On February 5, 1994, a 120 mm Serb artillery shell exploded in the center of Sarajevo's Makale market, killing 68 civilians and wounding another 144 people. In response to Serbia's war crimes, the UN Security Council passed Resolution 824 in June 1993 that designated five safe zones for the Bosniaks in Sarajevo, and in Zepa and Gorazde in the Republika Srpska. The resolution also authorized a UN protection force (UNPROFOR) to protect civilians in these zones, but the UNPROFOR was not authorized to use force either to protect itself or the civilians that it was committed to protect. The Serbs disregarded thwe UN resolution and continued their cmpaign of ethnic cleansimg. In July 1995 the Scorpion Brigade of the Republik Srpska Army mssacured 8,373 Bosniak men and boys at Srebrenica; it was the worst genocide since

The next day, UN Secretary-General Boutros Boutros Ghali requested NATO to carry out air strikes against Serb artillery positions around Sarajevo that were bombarding civilian targets in Sarajevo. Operation Deliberate Force conducted 3,515 sorties by US and British aircraft against Serb positions in Bosnia until December 1995. Operation Deliberate Force was a major factor in convincing Slobodan Milosevic, Serbia's president, to participate in the Dayton Peace Talks.

Negotiations to settle the conflict were held at Wright-Patterson Air Force Base at Dayton, Ohio, in November 1995 under the leadership of US Secretary of State Warren Christopher and Assistant Secretary of State Richard Holbrooke. The warring parties were brought together only after enormous pressure was applied. What resulted was a deeply flawed compromise that, though ending the fighting in Bosnia, created all the conditions for renewal of the conflict.

The Dayton Peace Agreement established a single state of Bosnia-Herzegovina, which consisted of two semi-independent states: the Federation of Bosnia and Herzegovina, inhabited mostly by Bosniaks and Croats, and the Serb-dominated Republika Srpska, each with its own government and army, which were later combined into the Bosnian Army with three brigades, each with three ethnically based battalions that later fought with the multinational forces in Iraq until December 2008.[22] Each government controlled taxation and education within its territory and had its own police force and its own foreign policy. The Dayton accords also created a NATO-led stabilization force (SFOR). Thirty-six countries, led by the United States and backed by NATO, contributed more than 60,000 troops to enforce the Dayton agreement. SFOR remained in Bosnia for more than ten years.

Bosnia-Herzegovina was to be governed by the same legislature and presidency, but the presidency was a triumvirate, whose three members came from the three ethnic groups, and each had a veto power over legislation. Ethnic quotas were established at all levels of government, and school classes were divided according to the ethnicity of their parents. The government was bloated with 160 triplicate ministries and a public sector that consumed nearly half of the country's GDP. The peace was maintained for more than a decade by the the robust NATO SFOR. Power was decentralized so that each ethnic group could rule itself and each had its owwn government, its own police force and educational system. Ethnic quotas were esstablished for all government bodies. It was not suprising that the system resulted in a political gridlock and corruption and economic growth was stunted.[23]

Dayton created the Peace Implementation Council (PIC) consisting of 55 countries and international organizations. Over $14 billion in foreign assistance flowed into Bosnia between 1996 and 2007.

In 2000, the countries in the former Yugoslavia were offered an opportunity to apply for membership in the European Union (EU) and NATO on the condition that they develop institutions that would facilitate their integration into those organizations. In both Brussels and Washington, it was hoped that the allure of membership in these institutions would motivate the Bosnians to overcome their differences, but this was not enough of an incentive to cause the ethnic leaders to relinquish the prerogatives they had acquired under the Dayton accords.[24]

At the same time, the international community initiated an effort to help the Boisnians revise their constitution to eliminate its flaws. The leading parties developed a series of constitutional reforms to strengthen the central government's council of ministers, but the initiative broke down when the Serbs and the leaders of the Croatian Democratic Union refused to accept much as the EU tried to push Croatia toward membership, Croatia's leaders refused to relinquish their prerogatives. Since the European financial crisis, EU membership has lost much of the allure that it night once have have had.[28] Today, an uneasy peace pervades the region and the civil institutions remain divided and inefficient, as they have for the past 16 years.

Mediation

Although reconciliation between hostile groups was achieved in the Robber's Cave experiment and between the belligerents after World War II mostly by ttheir own efforts, success may not always be possible without the help of an outside party—a mediator—and sometimes even a skilled mediator cannot bring hostile partires together because there is no common ground between them that can be the basis of an agreement. In some cases, however, a mediator can achieve agreement between former belligerents that they are unable to achieve through their own efforts. Examples of effective mediation in international disputes include the 1978 Camp David Accords in which President Jimmy Carter brought Egypt and Israel into an agreement to normalize relations between them, the 1993 Oslo Accords between Israel and the Palestinian Liberation Organization (PLO), and the 1998 Good Friday Agreements between the Protestants and Catholics in Northern Ireland.

Most of the intractable conflicts in the world today, such as the Israeli-Palestinian conflict and the Basque insurgency in Spain, are based on long-held myths such as the Palestinians' "right of return" and that Jerusalem, including both the Jewish and Arab areas, is the "eternal, undivided" capital of Israel, even though Israeli Jews seldon venture into the Arab parts of East Jerusalem. Because such myths can never be challenged openly, it is essential that the mediator have private, or "backchannel," access to the principles of both parties in which such issues can be discussed confiden tially.[25]

Mediation is a form "of negotiation in which a third party helps the contending parties find a solution they cannot find by themselves."[26] In international affairs, the mediator is usually a third state that, more often than not, has an interest in resolving the conflict and can bring some leverage to the mediation process. The mediator's task is to understand the strategic positions of the parties and then to attempt to find or create a common ground between them that can be the basis for the resolution of the conflict, but sometimes such common ground does not exist or cannot be created, as in the Israel-Palestine conflict. Because most armed conflicts have deep emotional and cultural roots, the mediator must be able to show "tough love" and bring leverage to the conflict in the form of either positive rewards o r punitive sanctions as Peter Hain did in the Northern Ireland negotiations.

One of the myths associated with mediation is that the mediator must be an outside, impartial broker, yet it is very rare that an effective mediator would have such a distance from the conflict that his interests in the belligerents would be equal. Senator George Mitchell was not seen as a neutral between the Catholics and Protestant in Northern

Ireland when he was negotiating the Good Friday Accords. The deputy leader of the Ulster Unionist Party told the press "that putting [Mitchell] in charge of the talks 'was the equivalent of appointing an American Serb to preside over the future of Croatia. . . . [It] was a nonstarter.' "[27] Additionally, neither Richard Holbrooke nor Dennis Ross were regarded as neutral in their mediations of the Bosni an and the Israeli-Palestinian conflicts. What mattered was not neutrality but who could have the most influence on the parties and who could alter their behavior.[32] In both cases, it was the power and influence of the United States that brought pressure on the parties. In negotiating the Oslo Agreements, Norway could offer inducements, but it could not offer any meaningful penalties; only the United States could do that. The Norwegians were capable of a supportive or facilitating role but could not be the broker.[28]

Although finding a resolution to a conflict is not always possible, a mediator can sometimes arrange to limit or contain a conflict by such means as arranging a truce or cease-fire. Sometimes a successful outcome is to transform the conflict so that it can be resolved at a later time. Parties that are not yet ready to end a conflict can sometimes be induced into a cease-fire or a truce. The mediator can then work to preserve that cease-fire or truce. Perhaps circumstances will change so as to make negotiations for a final settlement possible.[29] The death of Yasser Arafat in 1994 could have been such an event.

The mediator must first fully understand the strategic situation of the conflict, the overt positions of the parties, and attempt to find their bottom lines. He must then determine if there is any common ground that can provide a basis for a settlement, and then he must try to expand on it. If the parties have themselves approached a potential mediator or have accepted his offer to mediate, it may be an indication that some common ground already exists and that the parties have a common stake in finding a solution.[29] In many cases, the basis for an agreement may be very narrow, such as a temporary crease-fire, but such a limited agreement can provide the basis for continuing the negotiating process.

An essential role of a mediator is to be able to sensitize each side to the other'sc oncerns and grievances. Usually the parties are so wraped up in their own concerns and grievances that they cannot see that the other side also has legitimate grievances and usually blames the other side for the conflict. If each side understands the grievances of its negotiing partner, it may be easier for them to work with the mediator to build bridges across what divides them. In discussions with the Syrians about the Golan Heights, Dennis Ross, the US mediator between Israelis and Palestinians, would first tell the Syrians that he understood why the Heights were so important to them and then would address Israel's concerns that would have to be satisfied if Syriawere to regain the Golan Heights.

One of the most important tasks of the mediator is to be able to develop outcomes that he can present to the parties that can be the basis for negotiation. Such proposals should be considered as drafts that the parties can mark up, that the mediator can carry to the other side. It is also useful to have multiple issues at play at the same time, so that if compromise cannot be found in one issue, there are others in which concessions can be made in order for one side to accept the other's position on the first issue.

The mediator must also be able to formulate solutions that fuzz over contentious issues. At one point in Dennis Ross's permanent-status negotiations with the Israeli president, Ariel Sharon told Ross, "We cannot accept what they want, and they cannot accept what we want." Ross replied, "What if we can fuzz the issues of borders by creating industrial zones or special economic zones in border areas that effectively put Israeli and Palestinian economic enterprises on both sides of what would be the border? What if, at the same time, we say that security would be handled in these areas by the Israelis? If we did, the very concept of borders would change."[37] In another example cited by Ross, British Prime Minister John Major, not Senator Mitchell, came up with the idea that Mitchell and his colleagues used for decom-missioning the IRA weapons as a precondition for the all-party negotiations in Northern Ireland, that the decommissioning would run parallel with, rather than precede, the negotiations.[31]

The techniques described above can be used by both mediators and negotiators. An effective negotiator should certainly attempt to understand his partner's positions, and the background and reasons for those positions; he should attempt to find his partner's bottom line and then attempt to find or create common ground as a basis for an agreement. It is also desirable to have more than one issue in play at the same time so that tradeoffs and compromises among them can be created. The negotiator should also attempt to build a situation of trust and confidentiality with his negotiating partner.

Chapter 12

War and Conflict in the Contemporary World

> Peace can be defined as security and the secure access to resources that are essential for living. A disruption of such resources could prove disruptive of peace.
>
> Regendra K. Pachauri[1]

In this chapter we investigate the nature of armed conflict and threats of war that exists in the contemporary world. Both China and Russia are attempting to extend their power and influence in their respective regions. Russia under its President Vladimir Putin is attempting to reestablish some semblance of the Soviet empire that existed before the fall of the Soviet Union in the 1980s, and China is attempting to establish its hegemony over the Western Pacific region and over East Asia at the expense of its neighbors, many of which have long established alliance relations with the United States.

In the Middle East and South Asia a long dormant conflict between Sunni and Shia Muslims and between conservative Muslim states and radical Jihadist groups such as Al Qaeda and the Islamic State have broken out into bitter wars. The western powers have been drawn into these conflicts because of the terrorists threats they pose to the west.

The conflicts in sub-Sahara Africa stem mostly from ancient tribal conflicts and religious conflicts between Christians and Muslims. Many of these conflicts have their origins in the legacies that the European colonists left behind when they abandoned their African colonies after World War II.

Figure 12.1 - China's Nine Dash Map

Rising China

Although China was not involved in any major armed conflicts at the time of this writing, China has engaged in provocative acts that could eventually result in armed conflict in East and South Asia and in the Western Pacific region. Beyond its shores, China has laid claim to many islands, reefs, sandbars and shoals which are also claimed by its maritime neighbors. In fact China has laid claim to the entire South China Sea. Oil and gas are believed to lay beneath many of these disputed waters which are also rich fishing grounds, and all the claimants are anxious to pursue oil and gas explorations in these disputed waters.

The Spratly Islands form an archipelago of 750 reefs, atolls, cays, and islands located 400 nautical miles (n.mi.) east of Vietnam's southern tip in the South China Sea as shown in Figure 12.1. The total dry land area is only 3.5 square n.mi., but the archipelago spreads over an area of 125,000 square n.mi. The Spratly Islands are believed to be rich in oil and gas and also contain rich fishing grounds. Various of the islands are claimed and occupied by Vietnam, China, Taiwan, the Philippines and Malaysia. In May 2011, Chinese naval vessels accompanying Chinese fishing vessels opened fire on four Vietnamese fishing vessels in the archipelago.

The Paracel Islands are another group of islands in the South China Sea claimed by China, Vietnam, and Taiwan. They consist of 30 islets, sandbanks, and reefs spread over 4,400 square n.mi. In July 2012, the Vietnam National Assembly passed legislation demarcating Vietnam's territorial waters to include both the Spratly and Paracel Islands. In response, China placed a military garrison and a "legislative assembly" on one of the islands. On another reef, the Scarborough Shoal, claimed by both the Philippines and China, China placed a rope across the entrance to the reef's lagoon and has kept patrol boats in the adjacent waters.

A small group of uninhabited islands in the East China Sea, known as the Senkaku Islands in Japan and as the Diaoyu Islands in China, are claimed by Japan, China and Taiwan, but they have been administered by Japan since 1972 after the United States gave up its occupation of the islands. Oil has been discovered beneath its surrounding waters.

In 2010, a Chinese fishing boat rammed into two Japanese Naval Self-Defense boats patrolling the Senkakus. Japan responded by detaining the captain of the Chinese boat. The Chinese government then detained four Japanese citizens and threatened to cut off its supply of rare earth minerals that are critical to Japan's electronic industry if Japan did not immediately release the captain. Japan released the captain three weeks after his detention, China then demanded compensation and an apology from Japan.

Japan responded by recalling its ambassador from Beijing over the island dispute and in August 2012, Japan arrested 124 Chinese citizens from Hong Kong, who had landed on the islands with the intention of claiming them for China. The arrests motivated massive anti-Japan demonstrations in China, and, in response, several Japanese companies closed six of their factories in China and told their Japanese employees to remain inside the factory buildings. The Japanese government then announced plans to buy the islands from their private Japanese owners, and China accused Japan of trying to "steal the islands." Japan then sent two maritime law enforcement ships to the islands, and China responded by sending six of its own patrol boats to the islands. Then, in early November 2012, the United States and Japan conducted a joint military exercise off Okinawa, but they canceled an amphibious landing exercise because they considered it to be too provocative. Since then Chinese coast guard boats and aircraft have continually patrolled the islands. In late September 2012 two Taiwanese fishing boats entered the disputed waters and the Japanese patrol boats attacked the Taiwanese vessels with water from their fire hoses. The Taiwanese boats returned the water "fire" setting off what might be called the Water Cannon War.

In January 2014, China declared a vast region of the East China Sea as a Chinese Air Defense Identification Zone (ADIZ) which overlaps the ADIZs of Korea and Japan and also covers the Senkaku/Diaoyu Islands. Aircraft entering an ADIZ are required to notify the owner of the zone. The United States has advised its commercial aircraft to notify China in addition to Japan and Korea, but refuses to have its military aircraft do so. On May 24, 2014 Chinese fighter jets flew dangerously close to two Japanese reconnaissance aircraft in the overlapping ADIZs in the East China Sea. Then on November 7, 2014, China and Japan issued a joint statement declaring, "Both parties recognize that they had different views as to the emergence of tense situations in recent years in the waters of the East China Sea, including those around the Senkaku/Diaoyu Islands and share the view that through dialogue and consultation, they would prevent the deterioration of the situation, establish a crisis management mechanism and avert the rise of unforeseen circumstances."

In May 2014, China installed a large oil rig in the South China Sea, 120 n.mi. off the coast of Vietnam, clearly within Vietnam's 200 n.mi. exclusive economic zone, but within an area also claimed by China. China deployed an armada of 80 coast guard and naval ships in the vicinity of the rig to protect it. Vietnam responded by moving 35 of its Sea Guard ships to the region to protest the Chinese actions. Vietnam then claimed that the Chinese had rammed their ships more than 171 times over a period of four days, and that the Chinese doused their ships with water from its water cannons. Vietnam responded by allowing its citizens to destroy scores of factories owned by Chinese, Taiwanese, Japanese, South Korean and Singaporean firms. Four people were killed and more than a hundred injured in the riots. China evacuated more than a thousand of its workers from Vietnam. The Vietnamese government, fearing reprisals from foreign investors, attempted to quell the riots. In a meeting of the Association of South East Asian Nations (ASEAN), in Burma, the member countries refused to give Vietnam any support because they feared reprisals from China and because of the destruction of their factories. All this occurred in spite of a 2011 understanding between China and Vietnam on a framework for discussing maritime issues to prevent them from turning into serious confrontations.

On May 25, 2014, a Chinese fishing boat rammed into a Vietnamese fishing boat about 17 n.mi. southwest of the Chinese rig and sank it. (Its ten crew members were rescued.) China unexpectedly began to move the rig out of the Vietnamese 200 n.mi. exclusive economic zone on July 15, toward the Chinese island of Hainan saying that the rig had completed its exploratory work. China then announced that it planned to place four more rigs in the South China Sea.

In early 2014 the Chinese began dredging sand from the sea bottom and piling it around seven reefs in the Spratly Islands to create artificial islands. The reefs are also claimed by Philippians and Vietnam. At Fiery Cross Reef in the Spratly Islands, the Chinese have constructed a 10,000 foot runway capable of handling fighter and surveillance aircraft, built a harbor capable of handling military vessels and have deployed artillery pieces on the artificial island. In May, China ordered a US surveillance aircraft to leave the area.

China's Foreign Policy. China bases its claims of its "indisputable sovereignty" over all of these disputed territories and waters on its "Nine Dash Map" shown in Figure 12.1, that encompassed all of the South China Sea. The map was initially drawn in 1947 by the Kuomintang Government of the Republic of China to show China's "historical waters," and was subsequently adopted by China's communist government in 1949, and the claims were included in the Law of the People's Republic of China, China's constitution. The map also shows the 200 n.mi. exclusive economic zones of China, Vietnam, Malaysia, Brunei and the Philippines that were established by the 1982 United Nations Convention of the Law of the Sea (UNCOLS)* that has been ratified by 165 countries including China. However when China ratified the Convention, it attached reservations in which it claimed sovereignty over all the territories discussed above, and that it does not

* Prior to the 1982 UNCOLS, there was no accepted limit on the extent of a country's territorial waters. Some countries claimed three nautical miles, others 6 miles or 12 miles, and still others claimed that its territorial waters extended to an ocean depth of 200 meters (660 feet).

accept those provisions of the convention regarding the settlement of disputes concerning these territories.*[2]

China's foreign policy also has its roots in its "tribute system" that dates back to the early Chou period (1027-770 BCE) of Chinese history. According to the Chinese *Book of History,* the first Chou emperor received the "mandate from heaven" to rule all mankind, but the emperor had to be compassionate and benevolent to all foreigners. Foreigners were required to reciprocate this imperial benevolence by the humble submission to the emperor, which was to be demonstrated by the ceremonial presentation of gifts and by the kowtow ritual in which the supplicant kneels before the emperor on both knees and then prostrated himself three times with his forehead touching the ground. This callisthenic is repeated three times. As China's influence spread over East Asia, the tribute tradition developed into a mechanism by which neighboring tribes and nations could gain a place in the China-centric universe and became the basis of China's trade with its neighbors.[3] Thus China's neighbors acknowledged China's suzerainty over them, and today China's rulers continue to believe that all foreigners must acknowledge China's suzerainty in their relations with it. Furthermore China appears not to respect the accepted international law, but relies on its own concepts of international law. China's claims in the South China and East China Seas appear to be an expression of its claims of suzerainty in the region under its ancient tribute system.

A second source of China's foreign policy is the humiliation that China faced when it was forced to accept the unequal treaties in the 1850s after its defeat in the Opium Wars (1856-1860). These treaties opened new trading ports to foreigners. Previously China restricted its foreign trade to the single port of Canton (now Guangzhou). The treaties also allowed Christian missionaries to establish churches, hospitals and cemeteries at the new treaty ports; allowed foreigners to study the Mandarin language that was previously forbidden, and provided that foreigners could only be tried for crimes committed in China in

* Such reservations and unilateral statements which nations attach to treaties upon ratification are common. For example, when the United States ratified the Convention on the Prevention and Punishment of the Crime of Genocide, which provides that individuals, not states, can be punished for acts .of genocide, the United States attached a reservation to the Convention stating that US citizens may not be indicted for committing genocide without the consent of the US Government.

courts of their own nation—a doctrine known as extraterritoriality n The treaties also included a "most favored nation" clause, whereby any privilege granted to another country would automatically accrue to all other countries that were granted a "most favored nation" status.

Because of the humiliations that these treaties forced upon China, China adopted a policy of "wealth and power" under which it would first have to become wealthy, and then use that wealth to become powerful so that it could challenge the western powers and regain its rightful place in the world.[4] But China's quest for wealth was stymied for over a century by a series of setbacks and catastrophes that included invasion by Japan, World War II, the civil war between Chang Kai-shek's Kuomintang government and Mao Zedong's Communists and then Communist rule under Mao that included the disastrous Great Leap Forward and the Cultural Revolution. It was not until the, rise of Deng Xiaoping and the introduction of his Socialist Market Economy in 1978 that China's economic rise began. Since the 1980s, China has risen to become a major economic power whose interests reach far beyond its shores to the Western Hemisphere, Africa and Europe. China is now the second largest economy in the world after the United States, and is now vying to become the dominant power in Asia and the Western Pacific region. China's power rests mostly on its export trade to the United States and the European Union and its imports of raw materials and food mainly from Africa. China's exports exceed its imports and it holds over $ 3 trillion in foreign debt, including more than $ 1 trillion of US debt. Since about 2010, China has felt wealthy enough to challenge its neighbors to regain what it regards as its rightful place in the world.

Although China's maritime neighbors have tried to deal collectively with China over the disputed territories using the auspices of ASEAN, the arbitration services of the UNCLOS, which China had already rejected and the US offers to mediate the disputes, China has insisted on dealing with each nation separately and has strongly opposed their efforts to negotiate collectively and to engage outside parties.[8] At the 2010 ASEAN meeting, the Chinese Foreign Minister, Yang Jeichi, warned the other member states against working with outside powers to resolve their territorial disputes with China.

In early September 2012, when the US Secretary of State, Hilary Rodham Clinton met with Foreign Minister Yang in Beijing, Clinton called on China to discuss a "code of conduct" to address the

Figure 12.2 – United States Relations in the Western Pacific Region[5]

disputes in the region in the ASEAN. To which Minister Yang replied, "China has sovereignty over the islands of the South China Sea and the adjacent waters. There is plenty of historical and jurisprudential evidence for that." He insisted that China would deal with each of the other claimants independently, and that the United States should reconsider its strategy in Asia in the light of "the trends in our current e a a n d t h e r general wish of the countries of the region."[6] After North Korea sank a South Korean naval vessel in 2010, revealed its uranium enrichment facility and shelled a South Korean island, China twice warned both the United States and South Korea against conducting naval exercises in international waters near China.

United State–China Relations. Before China's rise, the United States was the dominate power in the Western Pacific region since it had defeated Spain at the Battle of Manila Bay in May 1898 during the

Spanish American War. Since World War II, the Korean and Vietnam wars, the United States has signed mutual defense treaties and agreements with Japan, the Philippines, Korea, Taiwan, and Vietnam, and is obligated to assist these countries if they are attacked. These countries look to the United States as a bulwark against the growing hegemony of China so that conflict between the United States and China is a natural result of the events described above.

In January 2012, the United States announced that it intends to "rebalance towards the Asia-Pacific region. . . . Our relationships with Asian allies and key partners are critical to the future to the stability and growth of the region. We will emphasize our existing alliances, which provide a vital foundation for Asia-Pacific security. We will also expand our networks of cooperation with emerging partners throughout the Asia-Pacific to ensure collective and capacity for securing common interests."[7]

The new policy has been called the US pivot to Asia and the Western Pacific, and it represents a major change in emphasis of US foreign and defense policy. Beijing, however, has reacted to the new policy as a US attempt to contain China in the same sense that the US policy toward the Soviet Union during the Cold War was to contain Soviet expansion. On several occasions, however, US officials have said that it is not the policy of the United States to contain China. In September 2012, the US Secretary of Defense, Leon Panetta, said in Beijing, "Our rebalance to the Asia- Pacific region is not an attempt to contain China. It is an attempt to engage China and expand its role in the Pacific." And in June 2014, President Obama said in the Philippines, "Our goal is not to contain China. Our goal is to make sure that international rules and norms are respected, and that includes . . . maritime disputes." But given the different approaches of the United States and China to their international relations, China sees the motivations behind US actions and statements differently than the United States has intended. China sees the US intention to enforce "international norms" as an attempt to use these norms to contain China.

In February 2014, Admiral Jonathan Greenert, US Chief of Naval Operations, said that the United States would fulfill its obligations under its mutual defense treaty with the Philippines to help the Philippines if it came under attack, to which a spokesman for the Chinese foreign ministry responded, "The United States, which is not a party concerned in the South China Sea dispute, should honor its commitments of not taking sides on the territorial sovereignty issue. It should be discreet in word and deed and do more for peace and stability of the Asia-Pacific, rather than the opposite." A day later, Secretary of State John Kerry said, "With respect to the South China Sea, it is important to resolve these differences in a peaceful, nonconfrontational way that honors the [UN Convention on the] Law of the Sea and honors the rule of law itself."[8] But China has rejected using the mediation provisions of the convention and has said that it will deal with each country separately on a bilateral basis.

At a security conference in Singapore in late May 2014, the US Defense Secretary, Chuck Hagel, accused China of intimidation and coercion. He told the conference that China had called the South China Sea "a sea of friendship, peace and cooperation. . . . But in recent months China has undertaken destabilizing, unilateral actions asserting its claims on the South China Sea." Earlier, the Chinese President Xi Jinping told the Conference on Interaction and Confidence Building Measures in Asia, "We need to innovate our security concepts, establish a new regional security cooperation architecture and jointly build a shared win-win road for Asian security."[9] The Conference included China, Russia and other Asian nations, but not the United States. At another conference in Beijing, Admiral Sun Jainguo, the deputy chief of the General Staff of the Peoples Liberation Army described the US alliance system in Asia as "as antiquated relic of the Cold War that should be replaced by an Asia-centric security architecture."[10] The above discussion is an example of China's assertiveness, and the difficulties that it has with its neighbors and with the United States.

Although a military conflict between the United States and China is a possibility, it is vital that such a conflict be avoided and both countries appear to understand this, but their relationship is plagued by some fundamental misunderstandings that have resulted in what has been called "strategic distrust." In 2012, the Brookings Institution in Washington initiated a project entitled "Addressing U.S.-China Strategic Distrust" in which a Chinese scholar Wang Jisi and a US China scholar, Kenneth Lieberthal, addressed each side's misunderstandings of the other.[11] According to their analysis, both Beijing and Washington appear to want to build a constructive partnership for the long term. Their current relationship is a mature one, and each understands the other's positions on all the major issues and has extensive dealings with the other at many levels. Their leaders meet frequently, and each year, and there are more than sixty meetings between various agencies of the two governments, including their military establishments. Nevertheless there appears to be a significant amount of misunderstanding and distrust on both sides regarding the other's long-term intentions that could produce attitudes and actions that would feed upon themselves to increase distrust and intensify misunderstandings regarding the intentions of the other.

China's hostility toward and distrust of the United States has existed since the creation of the People's Republic of China (PRC) in 1949. During the 1950s and 1960s, China's leaders viewed the United States as a ferocious imperialist power and the gravest political and military threat to China. China's leaders believe that since 2008, China has become a first class power but that the United States has tried to prevent China from achieving that status. They also believe that the United States is not trustworthy, that it is trying to divide and westernize China, and that the ultimate goal of the United States is to maintain its global hegemony and prevent emerging powers such as China from achieving their goals and enhancing their status. China's leaders believe that the surveillance activities along China's borders by the US Navy and Air Force are pernicious and disregard China's interests. Finally, China believes that the United States is a declining power and that it is "on the wrong side of history."

In the economic sphere, China believes that the United States is using Chinese–U.S. economic frictions as a scapegoat for its own economic failures. China sees the United States as setting up political barriers to Chinese investments in the United States and that American pressures on China to revalue its currency are a high-handed, unreasonable way to serve US interests at China's expense. China has also become concerned about the wisdom of holding so much US debt in US Treasury paper because of the falling value of the US dollar. China also sees US actions in the Middle East as an attempt to control global oil supplies.

The US view of China is somewhat different. American leaders expect Beijing to adopt policies that will lead to a cooperative long term relationship that includes efforts to reduce conflicts, to cooperate in providing regional and global public goods such as maritime security, lowering greenhouse gas emissions, and to provide mutual benefits for both countries. Although some interests of the two sides will continue to clash and their very different cultures will give rise to different viewpoints on many issues, both sides can seek win-win outcomes on many of these issues, but American leaders see evidence that China still views itself as Number Two and the United States as Number One and that the United States is attempting to restrain China's rise. American leaders also believe that China is trying to displace America's dominant position in the Pacific region, and that China's leaders view the U.S.–

China relationship as a zero-sum game—we win, you lose—whereas US leaders believe that it is possible to achieve a long term constructive win-win relationship between the two powers, though America's leaders realize that that these aspirations might not be realized. Thus, although they will continue to try to achieve mutually favorable outcomes, US leaders recognize that it is necessary to be able to cope with the possibility that in the future these aspirations may move in different directions.

US leaders observe that the Chinese military is developing force projection capabilities in the Western Pacific that will eventually threaten the US capabilities throughout the region, and China has not clearly articulated its strategies in the region. American leaders also believe that there is too little interaction between the US and Chinese militaries to provide America with credible assurances that China will not threaten US capabilities for protecting its alliances and its broader interests in the region. The United States views access to waters beyond China's territorial waters as vital to its own security and to that of its friends and allies; therefore, the United States has undertaken contingency planning and acquisition programs to ensure that China cannot deny it access to the region in times of conflict.

American leaders are also concerned about the deep cultural differences between China and the West. The Chinese political system seems impenetrable and US leaders have little understanding of its internal workings. What appears to be a seamless web of the Politburo Standing Committee policy may actually be the uncoordinated initiatives of several ministries. Relations between the politburo and the Chinese military are murky, and the extent to which the politburo is in charge of the military or to which the military acts independently of the politburo is unclear. China's failure to carry out its commitments to the United States—e.g., on the protection of intellectual property rights—tends to be seen as insincerity, whereas it may be because of the limited authority of the central government.

China is a mature power, and given the size of China's economy and its foreign exchange reserves, US leaders expect China to play a constructive role in reinforcing global norms and regimes regarding such things as nuclear nonproliferation and climate change but they worry about the strength of China's commitment to these norms. Many Americans believe that China is taking advantage of America's economic and political difficulties and is actively working to prevent America's recovery. Despite China's laws regarding respect for foreign intellectual property rights and its joining the relevant international conventions, the ongoing theft of US intellectual property creates the impression that it is part of a broad policy of national development and defense strategies. The cyber attacks on US entities from within China reinforces these beliefs, yet in 2013, it became clear that the United States was engaged in similar attacks against Chinese entities.

China's actions to keep the value of its currency below its real market value so as to under price its exports on the world markets is viewed as a subsidy on Chinese exports and a tax on US exports to China, and as hurting US efforts to create domestic manufacturing jobs. China has also placed severe restrictions on the export of rare earth metals since it has become the source of 90 percent of the global supply of these metals, thus severely limiting America's electronic and clean-energy industries. US firms can gain access to these materials only if they moved their production facilities to China, thus putting their technologies at risk of being stolen

The authors of the Brookings paper point out that although distrust may unavoidable, it is vitally necessary that China and the United States work to limit the damage of mistrust to their interests, else an incident could trigger an unwanted political or military crisis. They point out that developing strategic trust may be difficult because the sources of the distrust are deep and multifaceted and are not well understood. China claims that its foreign and defense policies are defensive to sustain its economic growth, to contain destabilizing forces from abroad, to prevent territorial losses, and to reduce its neighbors' suspicions. But apart from this broad objective of sustaining its economic growth and improving the wellbeing of its citizens, China has had a great difficulty articulating a comprehensive and congruent set of foreign and defense policies. This may be because of the many actors that claim to have a voice in the formation of these policies.[12, 13]

The China-US Military Relationship. China has been increasing its military expenditures at a rate of more than 15 percent each year between 1990 and 2009, by 7.1 percent in 2010, and by 12.7 percent in 2011 to 601.1 billion Yuan, the equivalent of about $ 91.5 billion dollars.[18] However since the Chinese defense budget does not include many of the items that are included in the US defense budget, on an equivalent basis the Chinese defense budget is likely to be between two and four times its stated amount. The FY 2012 US defense budget was $ 553 billion, so that in real terms China's defense budget could be between 30 and 50 percent of the US defense budget. Nevertheless China has the second largest defense budget in the world, and could become a formidable military opponent for the United States. China's People's Liberation Army (PLA) includes all service branches and has about 2.3 million men and women on acive duty.

A future war between the United States and China is likely to be a maritime war. Currently the Unites States has a vast naval superiority over China. It has eleven carrier strike forces, five of which are deployed in the Pacific region: one permanently stationed at Yokosuka, Japan, one assigned to the Seventh Fleet in the Western Pacific, one assigned to the Fifth Fleet in Southwestern Asia, one in Everett Washington and one in San Diego. Although China's navy includes an extensive fleet of attack and ballistic missile submarines, frigates and destroyers, China has no capital ships such as cruisers and battleships, and China commissioned its first aircraft carrier in 2011 which is a refurbished Russian carrier. Its air force consists of 250,000 personnel and about 25,000 fighters and attack aircraft. The Second Artillery Corps, its missile and space organization, is believed to have between 100 and 400 nuclear capable missiles, some of which have an intercontinental range and are capable of reaching the continental United States. China recently announced tha it was developing multiple independently targeted reentry vehicles (MIRVs) for its ballistic missiles to counter the missile defenses that the U.S./ is planning to deploy in the area. And China has also developed an extensive cyber warfare capability.

In May 2015, the Chinese State Council released a Defense White Paper that laid out a new Chinese military strategy that shifts emphasis from a strong land army to greater naval power with force projection capabilities particularly in the South China Sea. The paper also emphasizes a stronger air force and improved missile forces. It notes the construction of two new lighthouses in the Spratly Islands noting its territorial claims in the region, that cyberwarfare as a "grave security threat." Without motioning the United States specifically, China perceives it as China's major opponent, blaming "some external countries" for "meddling in South China Sea affairs." Xu Guangyu, a retired major general, remarked that, "As China continues to rise, it has enormous interests around the globe that need protection, including investments, trade, energy, imports and the surging presence of Chinese living abroad." He noted that the document stressed Beijing's resolve to deter foreign aggression and to win a war started by others. "China will actively build up its military capability and deterrence, just to make sure no one dares to fight with us."

Relations between China and the United States present a great challenge for the leaders of both countries. In spite of the relationships' apparent deterioration in recent years, both sides appear to want to avoid open hostilities. In 2010, China's premier, Wen Jiabao, said of the US–China relationship, "Our common interests far outweigh our differences." Even though China has had disputes with the United States, Japan, Philippians and South Korea, it recognizes that these countries are among its top economic partners and that it is essential that they remain so if China's development is to continue. According to Wang Jisi, dean of the School of International Studies at Peking University, "The best China can do is to strengthen its economic ties with great powers while minimizing the likelihood of a military or political confrontation with them."[15]

However as long as China continues its belligerent stance towards its neighbors. the United States should be able to deter China from aggressive actions by maintaining a balance of military power in the region. Currently, the United States has an overwhelming superiority of naval power in the region, but over the long term, China can be expected to challenge that superiority, but the United States should be able to meet those challenges. Over the long term, it may be possible to implement either an explicit or implicit arms control regime to avoid a conventional arms race in the area. The states in the region may also be able to implement a regional defense alliance that, with the help of the United States, should be able to deter any Chinese aggression.

At an Asia Pacific Economic Cooperation summit in Beijing in November 2014, the United States and China signed agreements on four issues: specific goals to limit their greenhouse gas emissions; to lower tariffs on the information technology products of the two countries; to issue visas for up to ten years which allow for multiple reentries into each other's countries; and they agreed to notify each other of their military exercises in the South China Sea so as to avoid serious confrontations. Although these agreements may appear to have ushered in a new era of cooperation between the United States and China, China can be expected to pursue a pragmatic foreign policy of cooperating in areas and on issues that are to its advantage, while at the same time seeking advantages in the Asia Pacific region at the expense of its neighbors when the costs are low.

The New Russian Threat

A new threat of war and aggression has arisen in eastern Europe since Vladimir Putin returned to the Russian presidency in 2012. Since Mikhail Gorbachev dissolved the Soviet Union in 1991, Russians have felt that their country has been treated as a defeated nation of the Cold War. Most of its former colonies, including Poland, Czechoslovakia, Albania, Bulgaria, Hungary and the Baltic republics of Lithuania, Latvia, and Estonia have joined the NATO alliance, and Putin and many other Russians feel betrayed by NATO. They believe that Gorbachev had been promised that NATO would not expand beyond a united Germany,* but those former republics and colonies of the Soviet Union clambered to join NATO , and now NATO is on Russia's borders in Poland, Latvia, Lithuania and Estonia. According to Mr. Putin, "They have lied to us many times, made decisions behind our backs, placed us before an accomplished fact. This happened with NATO's expansion to the east, as well as the deployment of military infrastructure at our borders. They keep telling us the same thing: 'Well, this does not concern you.' That's easy to say."[16]

During his second term as president of the Russian Federation, Mr. Putin has claimed that Russia has a moral obligation to protect ethnic Russians where ever they live and to bring them back into the Russian fold, and he has used this as an excuse to justify Russia's invasions into the South Ossetia region of Georgia in 2008, and into Crimea and the eastern regions of Ukraine in 2014. Putin's goal appears to be to recreate a resemblance of the former Soviet empire, although he claims to only want to create a free trade customs union—the Eurasian Customs Union (ECU). At the time of this writing only Russia, Belarus and Kazakhstan have joined the ECU. Other countries in the region, including Azerbaijan, Moldova and Georgia seem to have cold feet. Georgia fought a war with Russia as recently as 2008, and Ukraine is in a serious conflict with Russia over its annexation of Crimea and its stealth invasion into eastern Ukraine. During his second

* A recent review of available documents, including those from the West German Foreign Ministry and those made public by Wikileaks indicate that no such promises were made, although the extent of NATO's expansion was discussed by Gorbachev and the US Secretary of State James Baker and other western and Russian Leaders in the early 1990s. During those meetings Baker reminded Gorbachev that under the 1975 Helsinki Final Act, each country was permitted to choose the military alliance it wished to join.[18]

term as president of the Russian Federation, Putin declared, "The collapse of the Soviet Union was the greatest geopolitical catastrophe of the century. Tens of million of our citizens found themselves outside the Russian Federation."[18] Putin has strenuously opposed every move that Ukraine has made to improve its relations with the West and has complained about the West's attempts to reorder the balance of power on the continent.

Putin has adopted a messianic, nationalistic view of Russia's place in the world and in history. He is fond of quoting Russian philosophers such as Ian Ilyin (1883-1954), Nikolai Berdyaev (1874-1948) and Vladimir Soloyov (1853-1900), and has assigned them as required reading to the Russian regional governors. Ilyn was a nationalistic Russian philosopher who wrote, "We trust and are confident that the hour will come when Russia will rise from disintegration and humiliation and begin an epoch of new development and greatness. . . The West transported this anti-Christian virus to Russia. Having lost our bond with God and the Christian tradition, mankind has been morally blinded, gripped by materialism, irrationalism and nihilism." Solovyov has argued that since Russia is located between the Catholic West and the non-Christian east, it has a historic mission to lead the way to human unification. Russia would transcend secularism and atheism and create a unified spiritual kingdom. And Berdyaev wrote, "The Russian messianic conception always exalted Russia as a country that would help solve the problems of humanity." According to David Brooks, these thinkers believe that "Russia is frequently seen as a besieged fortress. The West is thought to be rotten to the core and [morally] weak, yet so powerful that it can be blamed for everything that goes wrong. Russia has immeasurable spiritual potential yet is forever plagued by the lack of self-respect, lack of self-assertion and unmet potential."[21]

In Eastern Europe, Putin has used all the leverage he has to prevent the former Soviet republics from building ties with the west, including cutting off gas deliveries to Ukraine and western Europe, banning imports of Moldovan wine and Ukrainian, Polish and Lithuanian agricultural products. Russia, together with Armenia, Belarus, Kazakhstan, Kyrgyzstan and Tajikistan, all former members of the Soviet Union, have come together to form the Collective Security Treaty Organization that requires its members to assist any member that is attacked, which seems to be oriented towards religious extremist and terrorist organizations in Central Asia.

Figure 12.3 – Ukraine.[20]

The 2014 Pro-Russian Insurgency in Ukraine. Ukraine's population consists of two major ethnic groups: ethnic Ukrainians located mostly in the western and central parts of the country, and ethnic Russians in the eastern part of the country, although there has been much intermarriage and ethnic mixing among the two groups in the east, and the population of these oblasts (provinces) is heavily pro-Russian. The percentages of ethnic Russian speakers in each of Ukraine's 24 provinces is shown in Figure 12.4. The eastern provinces are heavily industrialized with coal mines and Soviet-era heavy industry, and a large portion of its product is used in the Russian defense sector.

From 2006 until 2008, Ukraine was ruled by a pro-European faction that had petitioned for membership in the European Union and in NATO, but in April 2008, Vladimir Putin, then president of Russia, declared at a NATO-Russian summit in Bucharest that if Ukraine joined NATO, Russia would consider annexing Crimea and the eastern part of Ukraine.[26] In 2010, Victor Yanukovych, a pro-Russian from eastern Ukraine, was elected president of Ukraine in the wake of disputes with Russia over Ukrainian debts for natural gas after Russia briefly shut off its gas supplies to Ukraine in 2006 and 2009. The Yanukovych government was plagued with corruption, mis-management, lack of economic growth, and devaluation of the country's currency. In an attempt to remedy these problems,

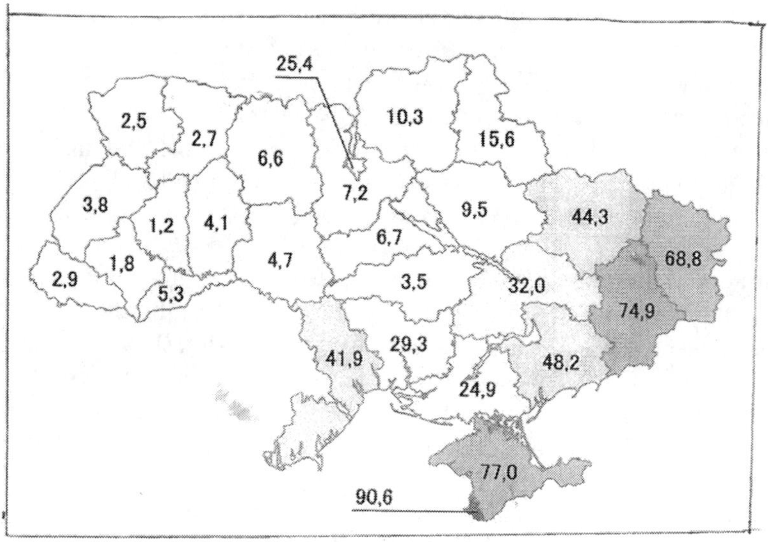

Figure 12.4 – Percentage of Native Russians and Other Russian Speakers in Ukraine.[22]

Yanukovych sought closer relations with both the European Union in the form of an association agreement and with Russia, but he refused to sign the EU association agreement because he thought its requirements were too austere for Ukraine. Instead, in December 2013, he signed the Ukrainian–Russian Action Plan, under which Russia would purchase $15 billion worth of Eurobonds from Ukraine and would reduce the price of Russia's natural gas by one-third. In exchange, Ukraine would cede the Kerch Peninsula on the eastern side of Crimea to Russia as shown in Figure 13.3. Putin said that he intended to build a bridge across the strait joining the Sea of Azov and the Black Sea, thus providing a direct access between Crimea and Russian territory.

The Ukrainian–Russian crisis began in November 2013 when demonstrations broke out in Euromaiden Square in Kiev, Ukraine's capital, in protest of Yanukovych's Ukraine–Russia Action Plan and to show support for closer ties to Western Europe. The protests evolved into calls for Yanukovych's resignation, and violence broke out in January 2014 after the government deployed troops to enforce its anti-protest laws. The protesters occupied government buildings in the center of Kiev, including that of the Ministry of Justice. Demonstrations from January 18 to 20 left 103 people dead and 1,419 injured from sniper bullets. On February 21, Ukraine's parliament in a vote of 328 to 0 with 129 abstentions determined that Yanukovych was unable to carry out his duties as president and set May 25 as the date to elect his replacement. Two days later, the parliament issued a warrant for Yanukovych's arrest and installed an interim government in what Russia called a coup d'état and declared the government to be fascist. Yanukovych fled to Russia.

Russian Annexation of Crimea. Before the Crimean Khanate was conquered by Russia in 1783 during the reign of Catherine the Great, it was a vassal state of the Ottoman Empire. Russia established its Black Sea Fleet in the same year and based it at the city of Sevastopol in Crimea as shown in Figure 12.3. Crimea became an autonomous region in the Soviet Union in 1921 after the Red Army took it from the pro-Tsarist White Army. In 1954 Chairman Nikita Khrushchev, who was a native of the region, transferred Crimea to the Soviet Socialist Republic of Ukraine, a symbolic gesture that seemed inconsequential at the time. Crimea regained its pre-1954 autonomy in the 1991 Crimean Sovereignty Referendum.

In 1997, Russia and Ukraine signed the Russian–Ukraine Partition Treaty, under which they agreed to divide the Black Sea Fleet between them. Russia was allowed to station 25,000 troops, 24 artillery units, 132 armored vehicles, and 22 military aircraft in Crimea. Both fleets were to be based at Sevastopol, and Russia agreed to pay Ukraine $ 98 million for the facilities they leased from Ukraine in Crimea. In April 2010, Russia and Ukraine agreed to extend the Russian lease between 25 and 30 years beyond the lease that would expire 2017. As part of the agreement, Russia agreed to supply gas to Ukraine at a discounted rate.

In February 2014, Russia began deploying additional troops into Crimea. Then on March 1, the Russian parliament voted to authorize President Putin to deploy Russian troops to Ukraine. The next day those troops already stationed in the Crimea mobilized without insignia on their uniforms, but using Russian marked vehicles took complete control of Crimea. Russia then held a referendum in Crimea, which was subsequently authorized by the Crimean parliament and the government of Sevastopol. The referendum asked the Crimean people whether they wanted to join Russia as a federal division or whether they wanted to restore the 1992 Crimean constitution and Crimea's status as a part of Ukraine. The 1992 Ukrainian constitution grants the Crimean parliament the power to establish diplomatic relations with other countries. The referendum was held on March 16 and 80 percent of Crimea's and Sevastopol's citizens voted to join the Russian Federation. The following day the Crimean parliament and the Sevastopol city council passed a joint Declaration of Independence of the Republic of Crimea and expressed their desire to join the Russian Federation. The very next day Russia and Crimea signed a treaty of accession of Crimea to the Russian Federation. The European Union, the United States, Canada, and several other nations declared the referendum to be illegal, and thirteen members of the UN Security Council voted in favor of a resolution that declared the referendum to be invalid, but Russia vetoed the resolution and China abstained. The UN General Assembly later adopted a resolution that declared the referendum to be invalid .

The Russian Stealth Invasion of Ukraine. During the Crimean crisis, Russia deployed between 40,000 and 50,000 troops, including armored units with tanks, along Ukraine's eastern border, claiming that they were conducting exercises. Then, several thousand Russian military intelligence officers crossed the porous border into Ukraine and began organizing ethnic Russians and pro-Russian Ukrainians for an insurgency against Ukraine's government. They wore military fatigues, carried Russian Kalashnikov assault rifles and some wore bulletproof vests. Several of the Russians admitted that they were from Chechnya and others said that they had fought in the Russian army during the Russian invasion of Afghanistan in the 19 80s. One of the leaders of the Russian insurgents was Igor Ivanovich Strelkov (the shooter), who was a colonel in the Russian Federal Security Service and a leader of the coup d'état in Crimea. Strelkov tried to bribe Ukrainian soldiers with cash. On March 24, Russia declared its readiness to intervene in eastern Ukraine to protect ethnic Russians there. It was apparent that the events described below were part of a playbook written beforehand in Moscow The troubles began on March 1 when pro-Russian activists occupied Ukrainian regional state administration buildings in several cities in eastern Ukraine, including Donetsk and Luhansk. But by the eleventh, the Ukrainian security service and local police had removed the insurgents from the buildings, and the next day violent clashes between pro- and anti-Russian factions broke out in Donetsk, and the local police appeared to be supporting the pro-Russian factions. Pro-

Russian insurgents forced about thirty pro-Ukranian activists onto a bus, then broke its windows and sprayed its interior with an irritant, forcing the group to leave the bus where they were then beaten and subjected to verbal abuse.

Between one and two thousand pro-Russian activists rallied in Donetsk demanding a Crimean style referendum to determine whether the oblast should join the Russian Federation. The referendum was scheduled to be held on May 11.The insurgents then stormed and took control of two floors of the regional state administration building in Donetsk and appealed to Putin to send peacekeeping troops to Donetsk. The next day the insurgents declared the formation of the Donetsk Peoples Republic.

On April 12, Russian agents with Kalashnikov assault rifles stormed and took control of the executive committee building in Slovyansk and police buildings across the oblast. Similar events occurred in the Luhansk, Kharkiv and Odessa oblasts. The Ukrainian military and police were unable to control the situation as the insurgents advanced their cause. In Moscow, the Russian Prime Minister Dmitry Medvedev declared, "Blood has been spilled and the threat of civil war looms."

Then on April 24, the pro-Russian insurgents declared the formation of the Luhansk Peoples Republic, demanded that Russian be made an official language, and that Kiev give amnesty to all the demonstrators. The pro-Russian insurgents in Donetsk and Luhansk proceeded with the May 11 referendums that asked, "Do you support the declaration of state independence of the Donetsk/Luhansk People's Republic?" In Donetsk, 89 percent of the voters voted in favor of independence in a turnout of 75 percent. In Luhansk, 95 percent voted for independence, and the turnout was 81 percent. Thus a large majority of the residents in both oblasts favored independence that presumably would result in joining the Russian Federation. However the United Nations, the United States, the European Union all declared the referendums to be illegal and Russia refused to recognized them. According to the Ukrainian constitution and its law on referendums, the entire country must vote on any territorial changes in Ukraine. Ukrainian forces then started to make probing attacks intro Slovyansk in the Donetsk oblast that was held by the separatists and established a blockade around the city. They gained control of all the check points held by the separatists, and control of the towns of Kramatosk and Andriivka, but the separatists later reoccupied Kramatosk. As the Ukrainian government initiated a new offensive, the separatists downed two government helicopters which resulted in the loss of life.

Putin withdrew the Russian troops from the Ukrainian border on May 7, and then on the tenth, Gozprom—Russia's natural gas company—increased Ukraine's bill for gas that had been already delivered by $ 2.2 billion and threatened to cutoff Ukraine's gas supply if it didn't pay.

Petro Poroshenko was elected president of Ukraine on May 25.. Poroshenko was an industrialist who made his fortune in confectionaries and was known as the "Chocolate King." He also had extensive business connections in Russia. Poroshenko had served in the government as Ukraine's Foreign Minister in 2009, and in 2012, he became the Minister Trade and Economic Development. One of his first actions after his inauguration was to intensify the campaign against the separatists. On June 26, he signed the association agreement with the European Union that Yanukovych had refused to sign in December 2013.

On June 3 Russia sent three obsolete T-64 tanks, several BM-21 multiple rocket launchers and other heavy military equipment into eastern Ukraine, and the next day the separatists shot down Ukraine military transport aircraft with a shoulder held heat seeking antiaircraft missile as the aircraft was approaching the Luhansk airport.

Poroshenko announced a unilateral ceasefire on June 10, and then on June 20, he offered to extend it by seven days. Although Putin accepted the ceasefire, the separatists refused to. During the ceasefire, Poroshenko reorganized the Ukrainian military's chain of command and the military gained more cohesion and discipline. By the middle of July, Ukrainian forces were able to rout the insurgents from Slovyansk and retook an important check point on the Russian border. In early August, Ukrainian forces had surrounded the cities of Donetsk and Luhansk. pounded the cities with heavy artillery barrages and gained control of significant parts of Luhansk after heavy street fighting. As a result the city was without water and electricity for more than two weeks.

In response to Ukraine's successes against the insurgents, Russia deployed about 20,000 troops along and across Ukraine's border. Russia then sent a convoy of about 200 trucks across the border to Luhansk which it said were for the humanitarian relief of the city, but Ukrainian officials were able to inspect only 23 of the trucks. Some of the trucks were only half full and one carried 20 tonnes of salt. The trucks were withdrawn the next day after delivering their cargoes. The contents of many of the trucks were unknown and may have contained weapons and ammunition or even Russian troops.

Then, beginning on the night of August 20, Russia launched a stealth invasion of Ukraine involving about 5,000 troops. A column of tanks, self-propelled artillery and modern SA-22 Russian air defense units began moving across the boarder into Ukraine. The movement of Russian men and equipment into Ukraine continued for several nights, but Russia denied that it had moved any personnel or equipment into Ukraine and claimed that any Russian soldiers that were in Ukraine were volunteers who had given up their leaves to assist the besieged pro-Russians in Ukraine. But NATO released satellite imagery showing the movement of Russian forces in eastern Ukraine and Ukraine's President Poroshenko ordered mandatory conscription to counter an "extremely difficult threat."

The pro-Russian insurgents with the help of Russian troops were successful in driving the Ukrainian troops out of Donetsk and Luhansk, and Russian forces then launched a new offensive against the city of Maruipol on the coast of the Sea of Azov in what appeared to be an attempt to open a land rout from the Russian border to Crimea.

On September 5, 2014, the presidents of Ukraine and Russia met with representatives of the pro-Russian insurgents in eastern Ukraine and of the Organization for Security and Cooperation in Europe (OSCE). They reached an agreement on a peace plan that included a ceasefire, the delivery of humanitarian aid to the cities of Luhansk and Donetsk, an exchange of prisoners, the removal of "illegal armed formations, military equipment and fighters and mercenaries" from Ukraine, and that Ukraine would pull its artillery units back beyond firing range of the cities occupied by pro-Russian and Russian forces. The ceasefire would be monitored by OSCE observers, but it did not include any political arrangements in eastern Ukraine. By September 10, Russia had withdrawn most of its troops from Ukraine[23] and on September 25, the Ukrainian Army and the insurgents exchanged prisoners.

But the ceasefire didn't hold and heavy fighting resumed with exchanges of artillery and tank fire in the Donetsk and Luhansk oblasts and at the Donetsk airport. At least 300 people were killed moved more troops and heavy weapons and equipment into eastern during the months of September and October. In November Russia Ukraine. But Putin forcefully denied that any of Russia's troops or equipment had crossed the border into Ukraine. The Donetsk airport became the scene of the heaviest fighting. By the end of January 2015 the airport had bencompletely destroyed. Its terminal building was in shambles and its runways had been totally destroyed, and t he Ukrainian Army was in chaotic retreat.

Economic Sanctions on Russia. In response to Russia's aggressive actions in Crimea and Ukraine, the United States, Canada and the European Union have imposed an increasingly severe list of economic sanctions on both individuals and commercial entities in both Russian and Crimea. The United States imposed its first round of sanctions i n March 6, 2014 restricting the travel of individuals deemed to have been responsible for violating the sovereignty and territorial integrity of Ukraine, and then expanded their list of sanctions on March 17 and 20 and again on April 26 to include 48 individuals and froze the assets of 17 Russian banks, energy companies and the investment accounts of four of Putin's closest associates. On July 17, the United States expanded its sanctions to include Russia's financial, energy and defense sectors and restricted the access of Russia's energy firms to western technology to access oil in the arctic, the deep sea and from shale. The European Union also sanctioned Russia for its activities in Crimea and in eastern Ukraine. In March, the European Union froze the funds and economic resources of designated persons, and in July, joined the United States in its sanctions against Russia's financial, energy and defense sectors and access to western energy extraction technology. Then beginning in the summer of 2014, the price of oil began to fall so that by the beginning of 2015, it had declined more than fifty percent, severely depriving Russia of income from its major export. The value of the Russian ruble fell dramatically causing Russian consumers to experience major inflation.

In August 2014, Russia announced its own sanctions against the west with a one year ban on the import of many food products, as long as they did not cause food prices in Russia to rise. Russian authorities also closed some McDonald's restaurants claiming that they did not meet Russia's sanitary standards. The western sanctions have caused food shortages, inflation and a fall in the value or the Russian ruble against other currencies.

In spite of the hardships which the sanctions have imposed on the Russian people, many continue their enthusiastic support of the government's actions in Ukraine. Putin's approval ratings have soared since Russia's annexation of Crimea and Russia's support of the Ukraine insurgency. Konstantin Somin of Higher School of

Economics in Moscow said, "The influence of economics as a whole has completely vanished. The country is on a holy mission. It's at war with the United States, so why would bother about the small battleground, the economy?"

Ukraine-NATO Relations. Although Ukraine is not a full member of NATO, it has had a close relationship with NATO since 1997 when Ukraine and NATO signed the Charter on a Distinctive Partnership and established the NATO Ukraine Commission, and again in 2002, NATO and Ukraine implemented the NATO Ukraine Action Plan for Ukraine to become a member of NATO. Ukraine has contributed forces to NATO's peace keeping operations in Kosovo, participated in NATO's training mission in Iraq, and Ukraine is a regular participant in NATO's military-to-military cooperation programs. But after Victor Yanukovych' election as Ukraine's president in February 2010, Ukraine's parliament voted that June to terminate Ukraine's integration into the Euro-Atlantic security system and membership in NATO from the country's national security strategy, thus placing Ukraine in the status of non-alignment between Russia and the West.

In September 2014, Ukraine's President Poroshenko attended the NATO Summit in Wales UK, and the alliance offered its strong support to Ukraine. Its Secretary General Anders Rasmussen declare, "Our support [of Ukraine] is concrete and tangible. . . . Ukraine has stood by NATO. Now in these difficult times, NATO stands by Ukraine." He said that NATO will provide Ukraine with about €15 million in assistance. and then said, "Russia must stop its aggressive actions against Ukraine, withdraw its thousands of troops from Ukraine and the border regions, and stop supporting the separatists in Ukraine."[24] Yet in spite of these brave words, NATO has not given Ukraine any tangible military support in its battle against Russia's offensive in eastern Ukraine. Indeed NATO has refused to give Ukraine any offensive weapons, fearing that Russia would provide even more weapons to the insurgents. NATO was originally formed to deter Soviet aggression against Western Europe, and to fight if Russia attacked western Europe. Most of the aid that has been provided by the west is non-lethal equipment such as helmets and communication equipment. The U.S. Congress has authorized $ 350 million in defensive weapons such as antitank weapons and artillery tracking radars, but the European countries have discouraged giving Ukraine any offensive weapons.

A New Peace Plan In February 2015 the presidents of Russia, France, the German Chancellor and representatives of the OSCE met in Minsk, Belarus to negotiate a new ceasefire in eastern Ukraine. Under the terms of the agreement, the cease-fire took effect at midnight Saturday, February 14. The agreement required both sides to withdraw their heavy artillery fifteen miles from the ceasefire line to create a 30 mile buffer zone. The Kiev government agreed to give amnesty to the rebels. Pro-Ukrainian and separatist militias were disbanded. Ukraine will regain control of its border with Russia at the end of 2015 if the government makes political concessions including constitutional amendments to decentralize its authority in the east. Ukraine agreed to resume paying pensions and other social benefits in eastern Ukraine. At the time of this writing in June 2015, the cease fire appears to be holding. Putin appears to have achieved his main objective of preventing Ukraine from joining the European Union and NATO/ A requirement for NATO membership is that an applicant is not involved in any territorial disputes. So long as Putin can keep the conflict in eastern Ukraine simmering in a "frozen conflict," Ukraine will not be eligible for NATO membership, and he can use economic pressure to prevent Ukraine from joining the EU. He has created such frozen conflicts in both Georgia and Moldova.

The New Islamic Threat

About twenty-three percent of the world's population, 1.7 billion people profess the Islamic faith, which is divided into two major branches, Sunni and Shiite. About 75 percent of t he world's Muslims, (believers in Islam) are Sunni and are located in Africa, Arabia, Palestine Turkey, Afghanistan, Pakistan, Central Asia, and Indonesia. The Shiites mostly occupy Iran, Iraq, Azerbaijan and Bahrain, with significant minorities in Syria and Lebanon—the so-called Shiite Crescent.

Since Islam's inception in the seventh century CE, Muslims have continued the warlike traditions of the Bedouin tribes of Arabia where Islam originated. During Muhammad's lifetime (circa 570–632 CE), Muslim Arabs conquered the Arabian Peninsula, and in the quarter century after his death, they extended their conquests across North Africa to Tripoli, north to include Syria, and east to the Afghan-Iran border, and over the next two centuries, they extended their conquests across all of North Africa, into Spain, and east to the Indus River and into Afghanistan.

Since Muhammad was both the political and military leader of his followers as well as their religious leader, there is no concept of the separation of the mosque and the state as there is in Western Civilization in the separation of church and state. Consequently, war, governance and religion are deeply intertwined in Islam, and wars are often fought with a religious justification, as they sometimes are in the west. The concept of the unity of civil and religious authority is contained in the Quran's mandate, "It is not up to any believer, man or woman, when Allah and his Apostle have passed judgment, to have any choice in their affairs."[25] Indeed, today most Muslims support religious governance as is evident of the election of Muhammad Morsi of the Muslim Brotherhood to the presidency of Egypt in August 2012, the election of Recep Erdogan of the religious oriented Justice and Development Party in Turkey as prime minister in 2003 and as president in 2014, and the theocratic government of Iran lead by the Supreme Leader and the Guardian Council, who are all senior Ayatollahs.

Yet in spite of this broad consensus among Muslims, there are at least four schools of thought regarding the relationship between Islam and governance. The traditionalists or conservatives believe that Islam is the divinely inspired blueprint for society and that the Shariah (Islamic law) is based on the Quran and on the Hadith or *Sunnah,* the teachings, sayings and religious habits of Muhammad, are divinely inspired and were fully developed during the early centuries of Islam. For the traditionalists, Islam is a closed system that governs all Muslims. It is based on and preserved in the medieval Islamic texts, and is vindicated by Muhammad's admonition in the Hadith, "The best of mankind is my generation, then those that follow them, and then those that follow them. Then there will come after them a people who will become avaricious, who will love gluttony and who will give witness before they are asked for it." For the traditionalists, Islam is an eternal blueprint for both personal and public life. Since Islamic law was divinely inspired, it is not the law that must be changed to conform to modern Islamic life, but society must "return to Islam" and conform to Allah's Law.[26]

The neo-traditionalists or progressives also maintain that the Quran, Hadith and Sunnah remain at the foundation of Muslim life, but they believe that for Muslim life to continue it must adapt to the modern world and incorporate western science and technology into its social and economic systems. The neo-traditionalists claim that they are not wedded to the classical synthesis as the traditionalists are, and that over the centuries the Shariah has incorporated some non-Islamic traditions, such as the stoning of women who have been raped that should be expunged from the law. The Muslim Brotherhood exemplifies the neo-traditionalist path, but it also claims the that the Quran and the Hadith should be interpreted to meet the requirements of the contemporary world, that Islam should be harmonized with the modern world. They also believe that Islam is the divinely inspired alternative to today's capitalism and communism.[27]

\In contrast to the conservatives and progressives are the contemporary Islamic reformers who are committed to "Islamic modernization" in which western technology and some western ideas are accepted into Islam, but they do not want to westernize Muslim society. They accept the early period of Islam as the normative ideal, but there are institutions and practices that should be changed in meet contemporary conditions. The reformers usually have been educated in western institutions but remain Muslims.[28]

Finally, the secularists advocate the complete separation of the mosque and the state, as in Western Civilization. They believe that the involvement of religion in political life is retrogressive and a major inhibition to the ability of the Muslim world to develop and compete in the modern world. They believe that religion should be limited to the private life, and that jihad is the lifelong struggle of the individual Muslim to be virtuous and to follow the straight path of Allah. To the traditionalists and neo-traditionalists, jihad is regarded as a community obligation and has several meanings: first, is the obligation to make the Muslim community more just and moral; second, is to spread Islam through preaching, teaching and through armed struggle; and third, is to protect Islam from outside efforts to subvert it.

In Egypt, Turkey and Pakistan, the military are the principle advocates of Islamic secularism and have attempted to suppress the conservatives and progressives as regressive and standing in the way of modernization. But in Egypt, the military secularists recognize that the soldiers and police serving under them are mostly conscripts who are also devout Muslims who view the ousting of a democratically elected president and the killing of civilians demonstrating against the military government is immoral and un- Islamic. Consequently, the Egyptian military's Department of Moral Affairs has enlisted clerics to make videos and issue fatwas to justify the killings on religious grounds. In one video a former mufti tells the soldiers, "Whenever somebody comes who tries to divide you, then kill them, whoever they are. . . . Even with the sanctity and greatness of blood, the Prophet permits us to fight this." In another video, a televangelist tells soldiers and police, "The day you wore that uniform and those boots and you made that salute and stood in your line—you are not doing a job [of killing civilian demonstrators] for a commander, you're working for God."[29]

The majority of Muslims today are either traditionalists or neo-traditionalist progressives. In the Egyptian parliamentary elections in August 2012, the Freedom and Justice Party and the New World Party, both parties of the Muslim Brotherhood won 37 percent of the of the seats in the parliament, and the Islamic Bloc of traditionalist parties won 27.8 percent of the seats, so that the traditionalists and neo-traditionalists together won 65 percent of the seats. The liberal, secularist Egyptian Bloc won only 8.9 percent of the seats, and the Coptic Christians, the rest.

Salafism. At the extreme right end of this spectrum of Islam is Salafism, a branch of Sunni Islam that advocates a literal interpretation of the Quran and of the teachings of Muhammad contained in the Hadith. The word *Salafi* means predecessors or ancestors in Arabic and refers to the founders of Islam during the first two generations after Muhammad. Salafism was founded in the latter half of the nineteenth century as a reaction to the spread of European liberalism and attempts to modernize the Muslim world. It advocates a violent jihad to establish an Islamic caliphate based on a strict application of the Shariah law, and that those who are not Salafists must either convert to Salafism or be killed. Salafism is an offshoot of Wahhabism which was founded by Muhammad bin Abd al-Wahhab (1703-1792) in the eighteenth century who preached a puritanical form of Islam and a return to the practices of the early Muslims. Wahhabism prohibits music, dancing, and the education of girls and prescribes beheadings of murderers, amputating the hands of thieves, and stoning to death female victims of rape because they are thought to have brought disgrace on their families.[30] Wahhabism and Salafism are especially strong in Saudi Arabia when 23 percent of the population profess either, in Qatar (47 percent) and in the United Arab Emigrates (UAE) (45 percent). Most Salafists consider the Wahhabism to be archaic and derogatory, whereas most Wahhabites consider Salafism to be too extreme, rigid and detached from mainstream Islam. Nevertheless, many wealthy persons in the Gulf states provide support for Salafism throughout the Muslim world. The extent to which the governments of Saudi Arabia, Qatar and the UAE funnel money through these individuals is not known, but the governments in these countries have appointed emissaries to their embassies in Muslim countries who proselytize for Salafism.

A revival of Islamic fundamentalism began in the 1970s and swept across the Muslim world to fill an ideological void left by the realization that the secularization that was taking place throughout the Muslim world could not fulfill spiritual aspirations of the Muslims. It was the result of the 1973 oil embargo and the subsequent four-fold increase in oil prices that exposed the western dependency on the Middle East. King Feisal of Saudi Arabia became a hero and aroused a new interest in the Saudi Arabian conservative brand of Islam, Wahhabism. Oil money flowed into the country that was used to build mosques and madrasses throughout the Muslim world. A second event was the Iranian revolution in which the secular government of the Pahlavi dynasty was replaced with the fundamentalist theocratic Islamic Republic of Ayatollah Khomeini, and the third event was the Soviet invasion and subsequent withdrawal of Afghanistan which precipitated a resurgence of political Islam.

Islam and the West. The Muslim conflict with the West dates back to the 11th century, when Pope Urban II raised an army of crusaders to drive the Muslims from the Holy Land. After seven major crusades and numerous minor ones over a period of 200 years, the Crusaders were unable to drive the Muslims from the Holy Land and were themselves forced out of Palestine.

In the twentieth century when petroleum from the Middle East became essential to fuel the industries and automobiles of the Western countries, Saudi Arabia and other Muslim countries obliged the West and sold their oil to the West on a strictly commercial basis, but when Western armies have invaded Muslim lands, they are seldom welcomed and are disdainfully referred to as crusaders. The principle objective of Osama bin Laden, the founder of al Qaeda, was to force the withdrawal of non-Islamist military forces and Western influences from the Muslim world, and to establish a worldwide caliphate with the strict application of Shariah law and Wahhabism or in some cases Salafism.

Al Qaeda. Although al Qaeda was founded in Afghanistan by Osama bin Laden to fight the Soviet occupation of Afghanistan (1979–1989), it has established franchises throughout the Muslim world, including Iraq, Somalia, Pakistan, Kashmir, Yemen, and the Islamic Maghreb (the Islamic far west in North Africa, west of Egypt). Since its founding, al Qaeda has carried out terrorist attacks throughout the world. In 1992 al Qaeda agents bombed the Movenpick Hotel and the parking lot of the Goldmohar Hotel in Aden, Yemen, where they believed US troops were staying en route to Somalia, but the troops were billeted in another hotel in Aden. In 1993, an al Qaeda agent, Ramzi Yousef drove a truck bomb into the parking garage of one of the World Trade Center towers in New York, hoping to cause the tower to collapse and to fall on the second tower, killing as many as 250,000 people, but although both towers swayed, they stood and the attack failed. In 1996, al Qaeda agents attempted to assassinate US President Bill Clinton when he was attending the Asia-Pacific Economic Cooperation conference in Manila in the Philippines. That same year, al Qaeda attacked the US embassies in Dar es Salaam, Tanzania, and in Nairobi, Ethiopia, with truck bombs, killing 223 people and injuring hundreds of others. The date of the bombing was the eighth anniversary of the arrival of US troops in Saudi Arabia to drive Iraqi troops out of Kuwait. In 2000, al Qaeda agents bombed the destroyer, the *USS Cole* in Somalia, killing 17 US servicemen and severely damaging the destroyer. Its most devastating attacks, however, were the September 11, 2001 attacks on the World Trade Center in New York and on the Pentagon outside Washington, DC, destroying both towers of the World Trade Center and destroying the west side of the Pentagon building, killing more than 3,000 people in New York and injuring thousands of others.

Although Osama bin Laden was killed by US Navy Seals in his compound in Abbottabad, Pakistan, in May 2011, al Qaeda franchises have remained active in Iraq, Syria, Somalia, Yemen, Kashmir, North Africa, and other places, and dozens of other Muslim extremist groups are also active in Africa, the Middle East and South Asia. Some of these groups are responsible for the attack on the US diplomatic mission in Benghazi, Libya, in September 2012, during which US Ambassador J. Christopher Stevens and three other US diplomats were killed, and for the hostage taking of workers at a natural-gas facility in Amenas, Algeria, in January 2013. Fifty-seven of the hostages, including three Americans, were killed when the Algerian Army attempted to rescue the hostages.

The Islamic State. The Islamic State is the most recent incarnation of the organization that began as the Organization for Monotheism and Jihad (JTJ), which was founded by Abu Musal al-Zarquwi who was born in one of the poorest neighborhoods of the town of Zarqu in Jordon. As a teenager he become heavily involved in petty crime and drug dealing. He dropped out of secondary school, drunk heavily and covered his body with tattoos. He was frequently arrested and during the time he spend in jails and prisons he came in contact with radical Islamic jihadists and developed a deep interest in radical Islam. In 1988, Zarquwi joined a Salafi mosque in Amman, Jordon where he received religious instruction. He gave up alcohol and removed his tattoos with acid. In 1989, Zarquwi traveled to Afghanistan with some of his fellow students to fight with the Taliban against the Russian invaders, but they arrived after the Russians had left. Zarquwi remained in Afghanistan and Pakistan and continued his religious studies in mosques and madrasses. He returned to Jordon in 1993, where he was arrested for planning a terrorist attack after explosives were found in his home. He was sentences to fifteen years in prison, but was pardoned in May 1999 by King Abdullah because of bad advice the king had received. Zarquwi then founded the Organization for Monotheism and Jihad (JTJ) with the objective of overthrowing Jordon's king, who Zarquwi thought to be un-Islamic.

Zarquwi returned to Afghanistan in August 1999 where he joined an al Qaeda camp in Herat and became its manager. Zarquwi had extensive interactions with Osama bin Laden and although they agreed that the ultimate goal of the jihad was the complete deliverance of Muslims from the western crusaders, they disagreed on the strategies for accomplishing this. Bin Laden with his ties to Saudi Arabia's political elite and his great wealth advocated a worldwide anti-imperialist jihad with the United States as the main enemy. Zarquwi, with his more limited background, thought that the jihad should focus on establishing a caliphate in his home country or in another country ruled by an apostate regime. Another difference between them was whereas bin Laden believed that non-believers included only non-Muslims, , Zarquwi believed that non-believers included all those who did not accept his extreme Salafi version of Islam.

After the 9/11 attacks in New York and Washington, Zarquwi went to Iraq in the summer of 2002, and founded the Organization for Jihad's Base in the Country of the Two Rivers, commonly known as Al Qaeda in Iraq, with a $200,000 grant from bin Laden, and began a series of terrorist attacks in Iraq against international and Shiite targets. During this period Zarquwi stayed in touch with bin Laden, and in December 2004, Zarquwi and his group pledged their allegiance to bin Laden, who named Zarquwi the emir of Al Qaeda in Iraq.

The goals of al Qaeda in Iraq were to force the withdrawal of the occupiers or "crusaders" from Iraq, to overthrow the Coalition Provisional Authority (CPA) in Iraq, to kill the collaborators with the occupiers and the CPA, to force the Shiites and their militias out of Iraq and to establish an Islamic caliphate based on the principles of Salafism.

Zarquwi developed a system of safe houses, weapon caches, military and tribal support networks in Iraq. Many of Al Qaeda in Iraq's members were former members of the Iraqi army that was disbanded by the CPA. Others included exiled Jordanians, Palestinians, Sunni Syrian jihadists and Islamists from Europe. Al Qaeda in Iraq financed itself by extortion, ransoms from kidnapping, underground trade in stolen antiquities and counterfeiting. They also received generous contributions from wealthy individuals and charities in the Gulf region including Saudi Arabia, Qatar and the United Arab Emirates.

Al Qaeda in Iraq claimed responsibility for: the attack on the UN Mission in Bagdad in 2004 in which Sargio Vieira de Mello, the head of the UN mission in Bagdad and a highly respected and well liked UN diplomat was killed; attacks on the International Red Cross and the International Red Crescent; suicide attacks in Bagdad's Green Zone; the destruction of the al-Askari Shiite shrine in Samarra; and, the 2003 attack on the Jordanian embassy in Bagdad in which 17 people were killed.

Zarquwi together with his spiritual advisor and his deputy were killed on June 7, 2006 in a joint US force operation. He was replaced as the leader of al Qaeda in Iraq by Abu Omar al-Baghdadi (a nom de guerre) who was killed in a US-Iraqi joint operation in April 2010. He was replaced by Abu Bakr al Baghdadi (also a nom de guerre). Over the years, 80 percent of al Qaeda in Iraq's leaders had been killed or captured, and by late 2012, its strength had dropped to about 2,500. Al Qaeda in Iraq retreated into eastern Syria in early 2013 where it was free from interference by the Syrian and Iraqi armies and could rebuild its strength. In July 2013, al Qaeda in Iraq conducted a raid at the Abu Ghraih prison in Iraq and freed 500 of its members including many of its senior leaders.

In April 2013, Baghdadi announced yet another name change of the organization. Now it became the Islamic State of al-Sham,* the Islamic State of Iraq and Syria (ISIS) or the Islamic State of Iraq and the Levant (ISIL). But in February 2014, Baghdadi published an online announcement that al Qaeda had disowned the ISIL and had expelled it from al Qaeda, and that the al Nusra Front had been designated as the sole franchise of al Qaeda in Syria.

In June 2014, the ISIS became simply the Islamic State (IS), a caliphate with Baghdadi the self-proclaimed caliph—a descendent of Muhammad and the spiritual, political and military leader of the Islamic State. Baghdadi declared that Muslims should unite under him as their caliph, and capture Rome so that they could "own the world." He said that the Islamic State would begin imposing Sharia duties on all its residents including the penalties prescribed in the Sharia, the establishment of Islamic courts, and begin levying of Islamic taxes. He declared that international boundaries between Islamic states were mute, because the Islamic State has established a single economic system and currency for all the territories under its control.[31]

By June 2014, the Islamic State had regained its strength which had grown to between 9,000 and 11,000 fighters including about 3,000 from Chechnya, France and Great Britain.[32] It launched a massive offensive in northern Iraq and seized control of Mosul, Iraq's second largest city, Falluja and Tikrit, and pushed within 120 km of Bagdad. In July, the Islamic State captured Iraq's largest oil field and Syria's gas field in Homs province.

In August 2014, the Islamic State and the Nusra Front invaded Lebanon and engaged the Lebanese Army at Arsel in a five day battle, but the Syrian Army was able to drive the rebel forces from Lebanon. The Islamic State offensive in the Sinjar region of Iraq forced between

---

* Al-Sham in Arabic refers to the greater Syria or to the Levant that extends from the Euphrates River to the Mediterranean coast and from the Tarus mountains in Turkey to the Arabian desert.

30,000 and 50,000 Yazidis* to flee into the Sinjar mountains, captured another 400 Yazidi women and threatened to kill them unless they converted to Salafism and agreed to marry its fighters. US President Obama authorized airdrops of food and water to the trapped Yazidis, and air strikes against the Islamic State fighters which allowed the Yazidis to escape from the mountains. Islamic State militants assassinated James Foley and Steven Sotloff by beheading them with knives. Both men were journalists working in Iraq and Syria. An Islamic State spokesperson said that they were killed in retaliation for US airstrikes on their forces in Iraq.

In June, Islamic State militants robbed Mosel's central bank of $ 429 million and took a large quantity of gold bullion. It also captured military weapons and equipment including, Russian SA-7 air defense systems, stinger shoulder launched antiaircraft missiles, antitank weapons, howitzers, Russian T-55 and T-72 tanks, multiple rocket launchers, and several UH-60 Blackhawk helicopters and cargo planes when it overran the Mosul airport in June 2014.

The Islamic State has developed a sophisticated and an effective recruiting and propaganda machine that uses contemporary social media and sate-of-the-art videos. Its messages are calibrated to appeal to different audiences. The message to western youth is "Be part of something bigger than yourself and be part of it now," and one Islamic State video features a Canadian recruit who urges North American Muslims to follow him and to bring their families, "You'd be very well taken care of here. Your families would live here in safety, just like how it is back home. You know we have expanses of territory here." A British Muslim fighter's recruiting message is, "Are you willing to sacrifice the fat job you've got, the big car, the family? Living in the West, I know how you feel—in the heart you feel depressed. The Prophet Muhammad said, 'The cure for depression is jihad.' "[33] The message to Muslim youth in the East is harsher. Arabic media portrays the jihad as the duty of all Muslims. It's videos show enemy corpses and groups of handcuffed prisoners being machine gunned down. In one video a fighter holds up his rifle and his passport and mocks his country, Bahrain, for threatening to withdraw his citizenship. He said, "Don't you know that you, your citizenship,

---

* The Yazidis are an indigenous Kurdish-speaking ethnic group in northern Iraq that has a unique religion that combines elements of mystical Sufism and Zoroastrianism, the ancient Persian religion.

your laws and your threats are under our feet. Don't you know that we are the soldiers of the Islamic State of Iraq and Syria and that our state will expand until it removes the thrones that you sold your religion for." In another video the message is, "Get out of the way or you will be crushed; join our caravan and make history." The videos and Twitter posts do not mention Israel or the United States. Instead the Islamic State is focused on the "near enemy," the Shiites in Syria, Iraq and Iran, "The struggle against the Americans and the Israeli is distant, not a priority. It has to await liberation at home."[34]

The Islamic State has issued two annual reports. The latest one released in March 31, 2014 and is over 400 pages long. Its contents are mostly a summary of the numbers of the kinds of attacks that it has carried out in Iraq by region. For example, it reports that it carried out 171 assassinations in Bagdad in 2012 and 95 in 2013; 450 improvised explosive attacks in the Anbar province of Iraq in 2012 and 617 attacks in 2013. In total there were 4,500 attacks in 2013 and 9,540 in 2014. The report shows that a vast majority of the attacks have taken place outside of the major cities and in rural areas, indicating Islamic State's objective to control territory.[35]

Sub-Saharan Africa

Nine (20 percent) of the ongoing wars in 2013 were in sub-Saharan Africa, and between 1998 and 2013, 58 percent of all the wars were in sub-Saharan Africa. A root cause of these wars is the legacy of European colonialism. The boundaries that the Europeans established between their colonies put ethnic groups and tribes, many of which were traditional antagonists, in the same colony so that when the colonies gained their independence during the 1950s and 1960s, these tribes and ethnic groups found themselves competing with one another for power.

From about 1870 until 1914, the European countries competed with one another for their African colonies in what was called "the scramble for power." The results of this competition are shown in Figure 12.6. Because the Europeans had neither the manpower nor the resources to administer their colonies, they turned to the native precolonial rulers to assist them. These local leaders then began to exploit their new positions to increase their influence in their own communities, and in many of the colonies, different ethnic groups and tribes competed for this power.

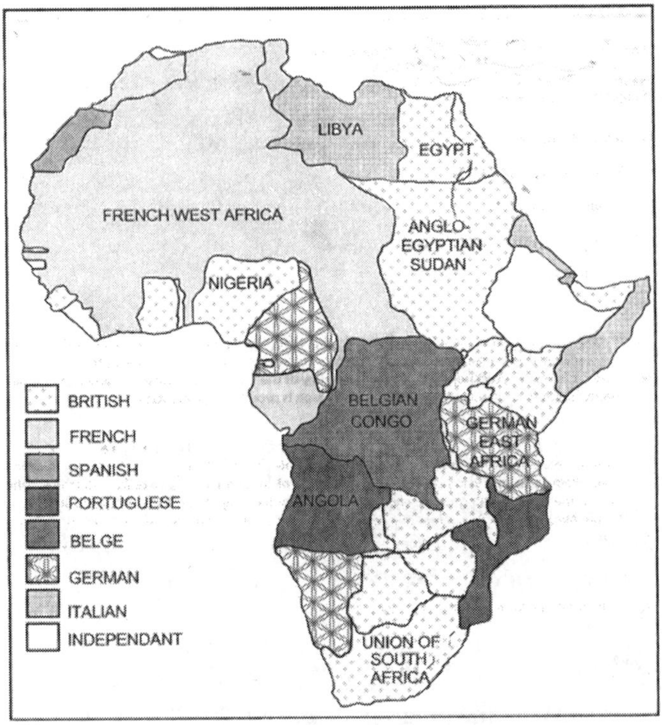

Figure 12.6. European African Colonies in 1914.

World War II left Europe severely weakened and more concerned with recovering from the ravages of the war. Since their colonies were no longer profitable, they had become burdens to their European rulers, and the colonies began clambering for their independence. Libya was the first to gain its independence from Italy in 1951. Tunisia and Morocco gained their independence from France in 1956, Ghana in the following year, and the rest gained their independence during the following decades. By 1980 there were 54 sovereign states in Africa, the vast majority of which were republics ruled by some form of presidential governance. Their borders followed those established for the convenience of their former colonial masters.

The Europeans had done little to prepare their colonies for independence, and in most cases, they had tried to imbue their colonists with their European cultural traditions, but in most cases, the colonists had been unable to assimilate those foreign traditions. The

British sent a few of their colonial youth to their universities and military schools in England, but when they returned home, they and the European cultures that they brought back with them were rejected by native populations. The French tried to inculcate the French culture and civilization into their colonies by insisting that they adopt the French language and culture and use French textbooks in their schools.

Nigeria is a typical example of the problems that the Europeans left. When Nigeria gained its independence from Great Britain in 1960, the country consisted of many tribes and ethnic groups but was dominated by three: the Muslim Hausa in the north, who formed the Nigerian Peoples Congress (NPC); the Igbo in the east, who were mostly Christians; and ethnic Cameroons, who formed the National Council of Nigerians and the Cameroons (NCNC). After independence, Nigeria was ruled by a coalition government of the three groups, but was dominated by the northern Islamic NPC. The first president was Jeremiah Obafemi Awolowo of the NPC. In 1961, the Cameroons in the south voted in a plebiscite to join the neighboring Republic of Cameroon, while those in the north opted to remain in Nigeria. This created a new ethnic imbalance, with the northern Muslim Hausa far outnumbering the southern Igbo Christians. In 1963, the country abandoned its British roots and declared itself a federal republic.

Because of the considerable rivalry between the ethnic groups, a civil war soon broke out between the Muslim north and the Christian south over accusations of corruption in the government and in elections. The result was a series of coup d'états and military dictatorships. The first, in January 1966, was led by a group of young Igbo military officers who murdered the prime minister of the western Igbo region and the premier of the Hausa northern region, but the coup leaders were unable to form an effective government and were forced to hand power over to the Nigerian army. This coup was followed by a second coup by a group of northern military officers who favored the NPC and installed Lt. Col. Yakubu Gowon as head of state. The coup was motivated by both ethnic and religious differences and resulted in increased tensions and violence between the north and west. The violence against the Igbo in the east resulted in their desire for autonomy, and in May 1967, the eastern region voted to declare its independence as a state called the Republic of Biafra, under the leadership of Lt. Col. Emeka Ojukwu. The Nigerian civil war began when the western and northern regions attacked Biafra in

July 1967. The 30-month war ended in January 1970 with the support of the British Army after a long siege of Biafra. The war left between one and three million dead from a combination of fighting, disease, and starvation.

The discovery of oil in the Niger Delta in the early 1970s created an economic and oil boom in Nigeria, but the military junta did nothing to improve the lives of the population; instead the military rulers became kleptomaniacs, pocketing the proceeds from the oil sales, making themselves very rich. A brief period of democratic rule in 1976 was followed by a series of coups and counter-coups until, in 1999, Olusegun Obasanjo, a former military dictator, was elected president of Nigeria and was reelected in 2003. His government was accused of being "unfair and unfree," but his government was able to show some improvements in tackling government corruption and in hastening development. Nevertheless, ethnic violence broke out in the Niger Delta over the disposition of the oil revenues. In 2007, Umaru Yar'Adua of the People's Democratic Party was elected president, but after his death in 2010, he was replaced by his vice president, Dr. Goodluck Jonathan, who led a moderate government that reduced corruption and provided services to his people, including clean water in the villages where lead released into the water supply from gold mines has killed hundreds of children.

In April 2014, Boko Haram, a militant Muslim group associated with the Islamic State and al Qaeda in the Islamic Maghreb kidnapped 275 secondary schoolgirls from a boarding school in Chibok in northeastern Nigeria. Fifty girls later escaped, but even with the help of western intelligence, including areal photographs, the Nigerian military has been unable to locate the girls who were taken into the bush. Boko Haram threatened to sell the girls into slavery unless the government releases their prisoners. Between 2009 and June 2014, Boko Haram murdered over 5,000 people in Nigeria.

In February 2015 Muhammadu Buham, a former military dictator in Nigerian, was elected the presidency to replace Jonathan Goodluck because of the corruption and inefficiency in Goodluck's government and because he had been unable to counter the Boko Harum and rescue the kidnapped schoolgirls.

Violence in East Africa

East Africa has been plagued with ethnic violence and warlordism for many years, including the 1994 ethnic conflict between the Hutus and Tutsis in Rwanda, the Dinka and the Nuer in South Sudan and the racial conflicts between the black farmers and Arab herders in the Darfur region of Sudan during the 1990s, but the most egregious violence is that due to the Lord's Resistance Army (LRA) in Uganda under the leadership of Joseph Kony.

The British, who ruled Uganda from the 1870s until its independence in 1966, divided Uganda along ethnic and linguistic lines with the Bantu speaking Buganda agriculturalists in the south and east, and the Nilotic speaking Acholi in the north, whose economy was based on a combination of hunting, herding, and agriculture. Following its independence in 1962, Uganda was ruled by a coalition government of the two ethnic groups, but it was overthrown by a coup in 1966. After a series of military coups, including that of Idi Amin, who ruled Uganda from 1971 to 1979. After Amin's ouster, the National Liberation Front formed an interim government that ruled Uganda until it was overthrown in a coup in 1986 by the National Resistance Army (NRA), lead by Paulo Mawanga. The NRA began a war of vengeance against the ethnic groups in the north of Uganda that included burning, looting, and killings. Their acts of terrorism led to the formation of several rebel groups, some including deserters from the Ugandan Army. Some of those groups made peace with Mawanga, but the rebels, who dominated the south of Uganda, continued to attack civilians in the north. In 1988, a civilian resistance movement was formed by Alice Lakwena, who believed that she was motivated by the Holy Spirit of God. Her followers used only stones and sticks as weapons, and she told them that they could protect themselves from bullets by covering themselves with shea nut oil and that they should never take cover or retreat from a battle. By August 1987, Lakwena's Holy Spirit Mobile Force won several victories in skirmishes against the Uganda Army and began a march on the capital, Kampala, but the movement was decisively defeated by government forces and Lakwena fled to Kenya.

The Lord's Resistance Army. Joseph Kony began life as a Catholic alter boy in his native village of Gulu in northern Uganda, but at the age of 15, he dropped out of school and became an apprentice to a village witvch doctor. In 1986 Kony became the leader of one of the many paramilitary groups that had sprung up after after the defeat of the Holy Spirits Movement. Kony's group, The Holy Spirit Salvation Army, was not perceived to be a threat to the National Resistance Army (NRA), but in 1986 the NRA merged with the Uganda People's Democratic Army (UPDA), which left many members of the UPDA dissatisfied, and some joined Kony's army, which he had renamed the Lord's Resistance Army (LRA). One such person was Odong Latek, a former brigadier in the UPDA, who convinced Kony to to use guerilla tactics instead of attacking in cross-shaped formations and sprinkling holy water. Kony's army included as many as 100,000 children who were abducted into the LRA. Some were forced to attack or kill family members or close friends and then to smear themselves with their victim's blood. They had only the choice of cooperating or being killed. Kony put his child soldiers in the front lines of his attacks because, although they were subject to high death rates, they could be easily replaced by raiding schools.

Kony began attacking villages both to demonstrate that the NRA was incapable of protecting the population against his attacks and to obtain supplies needed to support his army, which he forced villagers to transport and then held them captive for short periods. The new tactics yielded Kony's LRA several victories against the NRA in small-scale skirmishes, but the NRA retaliated against the LRA and almost decimated it. The LRA retaliated with a reign of terror, murdering NRA collaborators and mounting a terrorist campaign against the general population of Uganda.

In 1991, the NRA initiated Operation North, with the objective of destroying Kony and his LRA by cutting off its support from the population. Betty Oyella Bigombe, the Uganda minister in charge of Operation North and an Acholi, created Arrow Groups of villagers armed with only bows and arrows, who were easily overwhelmed by the LRA armed with modern weapons. Kony became convinced that the population had turned against him and initiated a reign of terror against the Acholi villages in the north who also had qualms against the occupying NRA.

After the failure of negotiations between the government and the LRA in 1993–1994, the LRA crossed the border into Sudan and established bases in southern Sudan with the permission of the government in Khartoum, as a response to Uganda's support of the Sudan Peoples Liberation Army, which was fighting for the independence of South Sudan. Kony began a campaign of terror against Acholi civilians in Uganad's north, mutilating them by cutting off their ears, lips and noses. In 1994 Kony began the mass abduction of chil dren into his army. The most infamous was the abduction of 139 girls in October 1986 from Saint Mary's College outside the town of Kerne in the center of Uganda. Because most of the LRA were children, an attack on the LRA was perceived as an atrocity against innocents. This created a moral dilemma between the necessity of defeating the LRA and the killing of children.

After the al Qaeda attacks in the United States on 9/11, relations between Sudan and Uganda improved and Uganda no longer gave support to the Sudan Popular Liberation Army (SPLA). In March 2002, the UPDA launched a massive offensive—Operation Iron Fist—against LRA bases in southern Sudan. After several months of fighting, the LRA crossed back into Uganda and began a reign of terror resulting in widespread suffering and displacement of the local populations. Beginning in 2004, a series of attempts to mediate peace between the warring factions by both local agencies and the United Nations were unable to halt to the violence. Finally in August 2006, the Ugandan government and the LRA signed a truce: the LRA agreed to leave Uganda and to deploy to two sites in southern Sudan. By June 2008, the LRA had acquired new weapons, recruited about 1,000 new members and was set to launch a new offensive, in which it murdered about 400 people in villages in the Democratic Republic of the Congo (DRC) and then continued its rampage in both the DRC and the Central African Republic. In response, the governments of Uganda, Sudan, and the DRC agreed on a plan to annihilate the LRA. In May 2012, Ugandan soldiers and an African Union brigade captured Major General Creaser Achellam, a senior LRA leader. In March 2013, the LRA consisted of about 2,000 fighters. Arrest warrants have been issued by the International Criminal Court in the Hague for Joseph Kony and four of his senior associates who remain at large.

Conclusion

In this chapter we have discussed conflict situations both within and among four civilizations, Chinese, Russian, Muslim and Western, each with its own cultures, values and interests, which are the foundation of their civilizations and are absolute and will not be compromised by their members. In the Western Pacific, before the middle of the nineteenth century, China had been the dominate power. It had developed its own system--the Tribute System--for dealing with others/ But when it began to decline after the Opium Wars, it adopted a policy that it first must become wealthy and then powerful in order to regain its rightful place in its universe to be able to deal with the western and other outside powers. From the end of the nineteenth century and especially after World War II, the United States became a major power in the Western Pacific region, setting up a potential conflict situation with China after it began to assert its power in the region in 2010. China has refused to accept western standards of international law and behavior as expressed in the UN Convention of the Law of the Sea.

Following the dissolution of the Soviet Union and its empire in 1989 and 1991, Russians believed that they have lost their rightful place in the world; that Russia is the chrysalis of Christianity and it is caught between a west that has lost its Christian faith and an atheistic east, and that its destiny is to revive the true Christianity. Russia further feels the need to unite the Slavic peoples into one nation wherever they are. It feels that it was wrongly defeated at the end of the Cold War, that its enemies have encroached to its borders, and that it must reunite the territories that it lost at then end of the twentieth century. Vladimir Putin is seen as a modern tsar who is able to restore Russia to its rightful place in the world and in history. Russia has infiltrated its agent-provocateurs into Ukraine, Romania, Moldavia, Georgia and Azerbaijan where they have stirred up pro-Russian sentiments, and the potential for lethal conflicts.

A minority of Sunni Muslims appear to be infected by a puritanical messianic complex of Salafism that demands that only Muslims of their brand of Islam are entitled to live and that all others must either convert to their faith or be killed. They feel that they are commanded by Allah to establish a universal caliphate governed by an extremist version of the Sharia Islamic law, and they are justified in using jihad to achieve their goals. The jihadists have incited armed conflict in Iraq, Syria and Palestine and in sub-Sahara Africa. The Muslim world is also torn between the religious traditionalists and the secularists, usually in the form of military junta, that are willing to use their military mussel to prevent religious traditionalists from rising to power. Military secularists have risen to power in Egypt,

Turky and Palestine.

Western civilization has long dominated the international system, first as a colonizer of nearly all of the non-western world, and then as originator of international law that is based on weatern concepts of justice and morality. It then imposed this systeem on the rest of the world. The west has also adopted a mssianic mission to spread democracy and othe \western values throughout the world. Althoug it has been somewhat successful in east Asia, it has been rejected i n most of the Muslim world, and where it has been tried in places like Iraq, Afghanistan, Turkey and Egypt, it has given rise to military juntas or sectarian dominated governments.as in sub-Saharan Africa are usually the result of ethnic xenophobia or what seems to be a senseless desire to kill with no political objective. The willingnes to kidnap and force them into military service is the height of moral depravity.

Chapter 13

Epilogue

> A statesman who yields to war fever must realize that once the signal is given, he is no longer the master of policy, but the slave of unforeseeable and uncontrollable events.
>
> Winston Churchill[1]

At the end of this study of why humans persist in their warlike habits, the reasons remain enigmatic. On the one hand, throughout history humans have sought ways to prevent war and have deplored its tragic consequences, but we humans have been unable desist from killing our fellow humans. We are eager to fight to defend our country, ethnic group or religion when they are attacked, and to aggressively expand them and force others to accept what we believe to be the true religion or form of government. We willingly risk our lives for our country, ethnic group or religion, and our country, ethnic group or religion willingly sacrifice their youth both to defend and to aggrandize its cause. Soldiers feel the thrill and excitement of killing their enemy. The love that soldiers feel towards their comrades and the risks they are willing to take for them are as real and as strong as the love that binds a man and a woman together in marriage.

Both of these impulses are deeply embedded in our genome and are the result of our evolutionary history. Our aggressive responses to threats to our societies and our xenophobia towards others who are not like us developed long long ago when our ancestors lived in small hunter-gatherer bands that competed with one another on the African savanna. To protect themselves a deep parochialism evolved that bound the members of these bands closely together, and a xenophobia towards outsiders also evolved.

Our aggressive instincts are part of our genetic inheritance, whereas our rational responses are part of our learned cultural inheritance which is passed from one generation to the next through teaching and learning.. Our inability to control our aggressive instincts attests to the strength and power of our inherited instincts.

We humans have adapted ourselves to live in very large societies or superorganisms that have many of the properties of a single organism, and we have become dependent on our superorganism—our nation, ethnic group or religion—to the extent that we cannot live without it. Thus we have become even more committed the survival and enhancement of our superorganism than we are to our own individual survival. Hence our willingness to risk our lives and fight for our superorganism. This commitment to our superorganism has engendered our nationalism and xenophobia and has been the cause of many wars.

Yet in spite of the terrible consequences of war, some wars have been beneficial to the human race. In the preceding chapters we have examined the role that war played in the rise of civilizations. Both the Roman and Ottoman empires which were formed through wars of conquest were able to establish an internal peace and prosperity that scarcely would have been possible without the wars that produced these empires. Some wars such as the Thirty Years War and World War II resolved issues that otherwise could not be resolved and new social environments emerged in their wake.

\Our ability to suppress these aggressive impulses and to seek rational solutions to our conflicts evolved much later and is located in a seperate part of our brain. Our aggressive instincts are located in our limbic system that was one of the earliest parts of the brain to evolve, whereas our ability to seek rational solutions to our problems is located in our neocortex that was the last part of our brain to evolve.

But even if one party to a conflict wishes to seek a peaceful solution, it is not always possible that the other side will accept a compromise solution. The 1938 Munich Conference is an example that sought to solve Hitler's aggressive threats against Czechoslovakia Hitler threatened to annex all of Czechoslovakia, but the other European powers proposed that Hitler could annex only Czechoslovakia's Sudetenland that contained a majority of ethnic Germans. Hitler agreed and the following day German troops moved into the Sudetenland. But six months later Hitler occupied the remaining parts of Czechoslovakia.

We have also seen how a country's populations enthusiastically supports its country's decision to go to war whether the country is attacked or if it is about to undertake an aggressive war, a condition called *rage militarie*, measning the patriotic enthusiasm for war. But as a war processes these same populations lose their enthusiasm as the costs of the war in terms of its casualities and national treasure become aparant.

Unfortunately, war seems to be a permanent condition of the our species that is divided into civilizations, nations, ethnic groups and religious sects that are so intolerant of each other that we are willing to kill those who are not like us. This intolerance seems to be one of the few social traits that is embedded in our genome. Almost all other of our social traits that are part of our cultures and are passed from one generation to the next by teaching and learning, not by genetic inheritance. Will we ever learn to live together peacefully and not to kill each

This conflict between our eagerness for war and our desire for peace is a conflict between our inherited instincts and our learned culture, between the devil within us and the "better angles of our nature."[2]

Appendix

Evolutionary Dynamics

> Nothing in biology makes sense except in the light of evolution.
>
> Theodosius Dobzhansky[1]

This appendix discusses the nature and processes of biological evolution. Darwin's Theory of Evolution by Natural Selection is the foundation of the modern theory of evolution, and indeed of all contemporary biology. Since Darwin's theory was published in 1859, it has been greatly developed and expanded because of the many discoveries concerning genetics and evolution that have been made since then, but Darwin's theory still remains at the foundation of all of contemporary biology.

Evolution by Natural Selection

Charles Darwin is one of the towering figures in science because he was both an astute observer and a theorist. As a young man of 22, he became the naturalist and captain's companion on the British Navy survey ship the *HMS Beagle.* From 1831 to 1836, the *Beagle* conducted a hydrographical survey of the coasts of South America, and then it continued its voyage around the world, stopping at the Galapagos Islands, Australia, and New Zealand. Darwin went ashore at all these places and acquired a vast collection of specimens, most of which had never been seen before, which he took back to England, where he began formulating his theory of evolution by natural selection, which he published in 1859 as *The Origin of Species by Means of Natural Selection.*[2]

Darwin's theory is based on four observations and assumptions:

- All species have the potential to reproduce so that the size of their populations would grow at an exponential rate;
- The resources for survivaal and reproducrion in any enviornment

are limited

- The ability to survive and reproduce varies among the members of a population of the same species; and
- Some of these abilities to survive and reproduce are heritable and are passed from one generation to the next.

From these observations, Darwin drew the following inferences:

- If more individuals are produced than the environment can support, there will be a struggle for survival among the members of the population;
- The outcome of this struggle for survival is not random. Those individuals that possess heritable traits that better suit them to the environment are likely to leave more offspring than the others;
- This unequal ability to survive and reproduce will result in a gradual change in a population, with favorable traits accumulating over many generations and unfavorable ones disappearing from the population and becoming extinct.

Natural selection is the result of the interaction between an environment and the differential ability of individuals in a population to survive and reproduce in that environment. The result of natural selection is the adaptation of a population to its environment, with favorable traits prevailing over less favorable ones, which may gradually become extinct.

It is important to note that evolution acts on populations and not on individuals. A population is defined as a group of individuals of the same species that occupy the same geographic area. Within a given population, there may be groups that leave more offspring with heritable traits than others. Thus, over time, a population will change. Populations are not necessarily isolated, and there need not be sharp boundaries between them.

A species is a group of interbreeding individuals that can produce viable offspring that can also survive and reproduce viable offspring. Although horses and donkeys can mate and bear mules as offspring, mules are incapable of reproducing, so horses and donkeys are therefore separate species.

Another important concept in biological evolution is fitness, which is the relative ability of the members of a population to survive and reproduce, and which is usually stated as the average number of offspring that different groups of a population produce.

Because natural selection operates on populations in specific environments, some populations of a species will endure while other populations of the same species in different environments will undergo evolutionary change so an existing species may continue to exist along with the new one. When the first hominids—the ancestors of the Homo genus—left the rain forest and began their evolutionary journey to become modern *Homo sapiens*, they left behind a species that continued to live in the rain forest and evolved into modern chimpanzees and bonobos. Thus, the diverse forms of life that exist today have evolved from ancestral species. The frequency of favorable traits gradually increases in each successive generation. Darwin incorporated this concept of gradualism into his theory; however even in Darwin's time, examples of rapid evolution were known.

An example of natural selection is the evolution of insecticide-resistant insect species. When a new type of insecticide is introduced, a small amount of the insecticide will usually kill 99 percent of the insects, but a small number of those insects will be immune to the insecticide and will survive and bear offspring that also are immune to the insecticide. Successive applications of the insecticide become less and less effective and kill fewer and fewer of the insects, until the insecticide becomes ineffective in controlling the pests. Over time, the insect population will consist only of those insects able to resist the insecticide. Natural selection has eliminated those insects from the population that are least able to survive and reproduce, and it allows those populations that have the greatest resistance to the insecticide to survive and reproduce. Herbert Spencer called this process the survival of the fittest.[3]

Mendel's Discovery of the Laws of Inheritance

The first modification to Darwin's theory was Gregor Mendel's discovery of the laws of inheritance. In 1857, Gregor Mendel, a monk in Bohemia (now the Czech Republic), began experimenting with pea plants to determine how their traits of flower color, seed and pod color and shape, and stem length are inherited by successive generations. Mendel's pea plants had either red or white flowers and he found that when he crossed plants with white flowers with plants with white flowers, their progeny all had white flowers, but when he crossed

	R	R		R	R		R	R		R	W		R	W		W	W
R	RR	RR	R	RR	RR	W	RW	RW	R	RR	RW	W	RW	WW	W	WW	WW
R	RR	RR	W	RW	RW	W	RW	RW	W	RW	WW	W	RW	WW	W	WW	WW
P(R) = 1				1			1			0.75			0.5				0
P(W) = 0				0			0			0.25			0.5				1

. Figure A.1. Punnett Squares for Six Types of Crosses for Two Alleles s from Each of Two Parents.

plants with red flowers with plants with white flowers, either all the progeny had red flowers or half were red and half were white. When he crossed red with red, either all their progeny had red flowers or, in some cases, ¾ of their progeny had red flowers, and ¼ had white flowers.

After crossing successive generations of pea plants, Mendel was able to sort out this confusion and concluded that there were two factors that determined the color of the flowers—a dominant factor and a recessive factor—and that each plant carried two factors, one of which is inherited from each of its parents. Pea plants are sexual and are either male or female. Thus, there are three possible combinations of the two factors: RR, WW, or RW. If the dominant red factor is present, only red flowers will result, and white flowers will result only if two recessive white factors are present.

Figure A.1 shows the Punnett squares for the six possible combinations of parental factors, the resulting combinations for the offspring, and the frequencies of the red and white flowers in the offspring of the two parents. The top row and first column of each square shows the two factors of each parent, and the four interior entries show the possible composition of the factors in the offspring. If the flowers have two identical factors, they are said to be *homozygote*, and if they carry two different factors, they are said to be *heterozygote*. In the first square, both parents are homozygote for red factor, and their progeny inherit only the red factor from each parent, so they are also homozygote.

In the second square, one parent is homozygote for the red factor, and the other is heterozygote and carries both the red and white factors, so that half their offspring inherit two red factors and are homozygote, and half inherit a red and a white factor flowers are red because all the flowers inherit the dominant red factor. The bottom two rows of Figure.

	W	W
R	RW	RW
R	RW	RW

→

	R	W
R	RR	RW
W	RW	WW

Figure A.2. Cross between Two Incomplete Dominant Homozygotes and between Their Progeny. (*RW* indicates a pink heterozygote.)

A.1 show the frequencies of the red and white flowers in the resulting populations

In the fourth square, both parents are heterozygote. One quarter of the offspring will be homozygote for red, one quarter will be homozygote for white, and half will be heterozygote and will have red flowers. Hence, three- quarters of the offspring will have red flowers and one quarter will have white flowers.

Incomplete Dominance. Mendel was lucky because he chose to work with pea plants whose factors are either dominant or recessive for the traits that he investigated, but for some genes, there is incomplete dominance, which in hybrids have an appearance somewhere between those of their parents. When a red snapdragon is crossed with a white one, the progeny will have pink flowers (RW), but if two pink snapdragons are crossed, only half of their progeny will be pink, one quarter will be white, and one quarter will be red. The Punnett square for incomplete dominance is shown in Figure A.2.

Polygenetic Inheritance. Many traits such as human skin color and height are determined by more than one gene and are called quantitative traits because the trait is continuous instead of having discrete values. Traits that have a continuous variation, such as height and intelligence, are usually quantitative traits, and their genes have an additive effect. Skin color, for example, is governed by at least three separately inherited genes. Random mating over several generations in a population in which a quantitative trait depends upon several genes usually results in a normal distribution for traits such as skin color, height, weight, and intelligence.

The Genetic Basis of Inheritance and Evolution

In 1865, Mendel published the results of his experiments in an obscure Czech journal.[4] and it was lost to the scientific community until it was discovered in 1900, 15 years after Mendel's death, when three botanists working independently reproduced Mendel's work. After searching the scientific literature, they found Mendel's paper and Mendel received the credit for discovering the "laws of inheritance." About the same time as the rediscovery of Mendel's paper, improved microscopes enabled scientists to see the chromosomes within the nuclei of cells, and a chromosomal theory of inheritance began to be developed. Early in the 20th century, Thomas Hunt Morgan at Columbia University, working with the fruit fly, *Drosophila,* was the first to advance the gene theory: that genes are the units of inheritance which are located on chromosomes, and that Mendel's factors are variants of a gene, that are now known as alleles.* It was soon discovered that chromosomes occur in pairs and that individuals of a sexually reproducing species inherit one set of chromosomes from each of its parents, thus confirming Mendel's laws of inheritance.

These discoveries resulted in the modern synthesis of evolutionary biology, which integrated the Mendelian genetics of inheritance with Darwin's theory of evolution by natural selection. The modern synthesis also emphasized the importance of populations as the units of evolution, the central role of natural selection in evolution, and the idea of gradualism to explain how large changes can result from the accumulation of many small changes accumulating over long periods of time. Although the modern synthesis remains the paradigm of biological evolution, discoveries over the past century have allele of a particular gene, the allele is said to be fixed in the population and the population is said to be homozygous with respect to that allele. In the contemporary theory of evolution, evolution is conceived as the change in the frequencies of alleles in the gene pool of a population.

The gene pool of a population is the total number of alleles present in the population. If all the members of a population all possess the same allele of a particular gene, the allele is said to be fixed in the population. In the contemporary theory of evolution, evolution conceived as a change in the frequencies of the alleles in the gene pool of the population.

* From allomorph, Greek for *different form.*

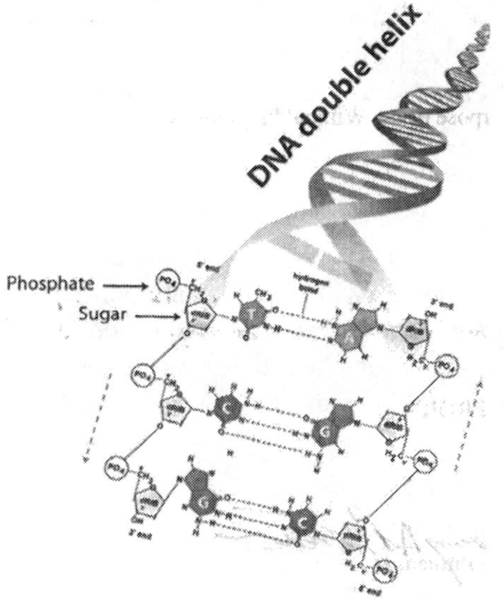

Figure A.3. The Double-Helix Structure of DNA[5]

DNA and the Genetic Code

In 1953, James Watson and Francis Crick, two graduate students at Cambridge University in England, published their famous paper describing the double-helix structure of deoxyribonucleic acid (DNA) the molecule that constitutes chromosomes. Chromosomes consist of two strands of DNA wound together in a double helix, as shown in Figure A.3. The two strands of DNA are linked by base pairs. Each strand consists of alternating phosphate and sugar groups. The sugar group is a five-carbon group (the pentagon shaped units in Figure A.3b called deoxyribose that are joined by a pair of bases (*AT, CG, GC*, or *TA*).

Four bases make up the genetic code, adenine (*A*), cytosine (*C*), guanine (*G*), and thymine (*T*). *A* and *T* are always linked together, as are *G* and A. A gene consists of a sequence of base pairs, and a triplet of three bases is a codon that is the code for one of the 20 amino acids, plus a start and three stop codes that signal the beginning and end of the gene. Because there are only 20 amino acids and three stop codes, and 64 possible arrangements of the four bases, there is considerable redundancy in the genetic code. (*GCU, GCC, GCA,* and *GCG* all code for the amino acid alanine.) A sequence of codons is translated into either a protein or an enzyme, which is a biological catalyst that regulates the rate of biological processes. In addition to genes that code for proteins and enzymes, there are other genes which control the expression of one or more coding genes and are called regulatory genes.

The Hardy-Weinberg Theorem

A useful concept that describes a population that does not evolve is the Hardy-Weinberg theorem, which was independently discovered in 1908 by the two men whose names the theorem bears. The theorem states that the frequency of the alleles in a gene pool will remain constant over generations unless one or more of the following conditions is violated:

- A large population so that random fluctuations will not change the allele frequencies;
- Random matting between male and females;
- No mutations of the genes in the gene pool;
- No immigration into the population to prevent the introduction of new genes;
- All members of the population have the same fitness.
- No genetic drift. (see below)

The Mechanisms of Evolution

If any of the five conditions of the Hardy-Weinberg theorem are violated, there will be changes in the allele frequencies in a population, and evolution may occur. The following paragraphs discuss the various ways in which evolution can occur.

Genetic Drift. If you were to flip a coin a thousand times, you would expect the result to be very close to 500 heads and 500 tails, but the result may depart slightly from this expected outcome and yield 497 heads and 503 tails. But if you flipped the coin only ten times, it wouldn't surprise you if the result was six heads and four tails, or three heads and seven tails. The difference between an expected result and the actual result is called a sampling error that usually occurs for small samples. Now consider that we have a population of 500 plants, whose gene pool has equal numbers of A_1 and A_2 alleles.

Because that each plant may produce hundreds of seeds, and we are drawing 500 pairs of the alleles from a very large gene pool. Since each seed has two alleles, there three possibilities—A_1A_1, A_2A_2, and A_1A_2—and the expected outcomes would be very close to 250 A_1A_1, 250 A_2A_2, and 500 A_1A_2 according to Mendel's laws of inheritance. Suppose, however, that we plant only ten seeds. It is highly unlikely that our sample would include exactly ten A_1 and ten A_2 alleles. The allele pairs of the ten seeds are more likely to be something like three A_1A_1, four A_2A_2, and three A_1A_2, or something like four A_1A_1, two A_2A_2, and four A_1A_2. In the first case, the ratio of A_1's to A_2's is 9:11, and in the second case, the ratio is 12:8. Now we select and plant ten seeds from the progeny of these plants. Because each plant produces many seeds, we are selecting the seeds from a large population of A_1's and A_2's, but the ratios have changed from the original ratio of 1:1, so the probabilities of drawing a specific number of an allele have also changed. If we repeat this process over many generations, the population of alleles will gradually drift so that one of the two alleles will become "fixed" in the population, the other allele disappears from the population, and the population become homozygous in one of the alleles. This process is called genetic drift. Sewall Wright showed that the probability that an allele will drift to fixation is equal to the allele's initial frequency in the parental population[6] and Motoo Kimura and Tomoko Ohta[7] have shown that the average number of generations for a mutant gene to reach fixation is approximately $4N$ generations, where N is the size of the population. Evolution by genetic drift is very unlikely in very large populations: For a population of a million and a generation time of 20 years, the average time to fixation would be 80 million years, much longer than the time from the hominin-chimpanzee speciation, which occurred about 6.5 million years ago. Thus, human evolution by genetic drift occurred when the human population was small.

The effect of genetic drift is to reduce the genetic diversity in a population so that it will be less responsive to environmental changes. Genetic drift can be a cause of evolution because it is a violation of the last of the Hardy- Weinberg conditions for a non-evolving population.

The Bottleneck Effect. Natural disasters such as earthquakes, droughts, floods, fires, disease, and the emigration of a small group from a large population may reduce the size of a population so that frequencies of the alleles in the resulting gene pool are different from those of the original population because of a sampling error. Some alleles may be eliminated from the population while the frequency of others may be increased. Genetic drift may further eliminate other alleles from the small population. The bottleneck effect can threaten an endangered species. For example, the cheetah was once widespread throughout Africa and Asia; however because of disease, human hunting, and droughts, the cheetah population has suffered a bottleneck effect and has become an endangered species.

The Founder Effect. When a small group emigrates from a large population to colonize a new location, it may suffer from both the bottleneck effect and genetic drift. The smaller the group, the greater the effect of both effects. The result is a population with less genetic diversity. An extreme example of the founder effect is when a single individual such as a seed is taken to a new environment, or when a single woman, for example, when mtEve, who lived about 190 thousand years ago, became the founding mother from whom all modern humans are believed to be descended.

One consequence of the founder effect is the propagation of genetic disorders. In 1814, 15 people migrated from England to the Tristan da Cunba Islands in the South Atlantic Ocean, midway between Africa and South America. One of the migrants carried a recessive allele for retinitis pigmentosa that causes progressive blindness in homozygotic individuals for the allele. Of the 240 descendants of the original settlers, in the 1960s, 4 suffered from retinitis pigmentosa, and 9 others were known to be heterozygotic carriers of the allele. The frequency of the allele is much higher in the islands than it is in the original population from which the population originated, so that according to Mendel's Law, the probability that a child of two heterozygotic individual will be homozygotic for the defective allele and have pigmentosa is 0.25, and half of the homozygote and a heterozygote will have pigmentosa.

Gene Flow. When a group of individuals migrates from one population into another and interbreeds with the host population, it may introduce new alleles into the receiving population, and over time, a new population may emerge. An example of gene flow is the immigration of Europeans, African blacks, Latinos, and Asians into the United States and their intermarriage with the descendants of the original settlers, so that today the United States is probably the most interbred, polyglot population in the world and over time can be expected to become a more homogenous mixed population. As the trend for humans to move across the world increases, gene flow is becoming an increasingly important agent of evolutionary change. After World War II, American service men and women married women and men of the countries in which they were stationed and brought them back to the United States, and their children inherited genes from both their parents. Increased gene flow and the homogenization of the world's population can be a major factor in reducing the frequency of wars, but it is likely to take many, many generations before this effect will be felt. A downside of such immigration is the introduction of new genetic diseases into a large population. Such gene flow can eventually change social attitudes towards minority groups and promote tolerance of minorities.

Mutations

Normally, there is enough genetic diversity in a population to allow it to adapt to a new environment; however occasionally, mutations in the genome occur that will facilitate environmental adaptations or produce novel traits that are beneficial to a population. Most mutations are either neutral or deleterious, resulting in genetic diseases such as cancer. Novel alleles can be created only through mutations. A genetic mutation is a change in the sequence of bases in the genome, and a mutation that changes a gene that is carried to offspring may change the gene pool of a population by adding a new allele to the gene pool. Mutations in a single gene are rare, and most are neutral and have no effect on the genome or on the individual. Most advanced species carry tens of thousands of genes—the human genome consists of about 21,000 genes—and many human populations are very large with tens to hundreds of millions of individuals, so that mutations are likely to occur frequently in the gene pool and the cumulative effect of a mutation over many generations can be significant. Most mutations occur during the process of chromosome copying, but most copying errors are corrected by proofreading and correcting mechanisms in the genome. Deleterious mutations in a gene generally result in a recessive allele that would have no effect unless the carrier mated with another carrier of the mutation.

Mutations that cause evolutionary changes must occur in either eggs and sperm (gametes) that are the carriers of genetic information from one generation to the next, or in structures called gametophores, in which eggs and sperm are produced. A mutation in the cell of a body organ may cause a disease such as cancer in that organ, but that mutation will not be passed on to offspring. However if the same mutation occurs in a gamete or gametophore, the carrier will not get the disease, but its descendants could.

Base-pair Substitutions. Point mutations are the simplest types of mutations. A base-pair mutation is the substitution of one base pair for another at a locus on a chromosome. There are four types of base pairs—*AT, TA, CG,* and *GC*—and these may be interchanged at a loci. Because most genes code for a protein or an enzyme, and there is a significant amount of redundancy in the genetic code (64 codons code for only 20 amino acids, the start and stop codons), thus many base-pair substitution mutations will have no effect and are called silent mutations. If a base-pair substitution codes for a different amino acid, it is called a missense mutation. Occasionally, a missense mutation results in a different amino acid that yields a protein with capabilities that enhance the fitness of a mutant individual and its progeny, but more often, such mutations are deleterious because they either reduce the fitness of the carrier or cause a deformity or a disease. There are also long stretches of DNA in a genome that are not known to have any functional use, where base-pair substitutions would be silent. Base-pair substitution that causes a codon to become a stop codon is called a non-sense mutation and results in a shortened, nonfunctional protein. If a base-pair substitution occurs in the *AUG* start codon, then the gene will not be transcribed and the protein will not be transcribed that could cause a malfunctioning of the cell.

An example of a base-pair substitution is the change of a single base pair in the gene for the human hemoglobin molecule that consists of 146 amino acids. The base-pair substitution in the sixth codon changed it from one for glutamate to valine. The mutation provides its carriers with an immunity to malaria and has spread throughout the human population in Africa so that approximately 25 percent of the West African population carries the mutation. In individuals who are homozygotic for the mutation, however their hemoglobin become sickle shaped and the ability of their hemoglobin to carry oxygen is significantly reduced and they suffer from sickle-cell anemia and usually die before the age of 50. Sickle-sell immunity to malaria is an example of natural selection, because heterozygotic carriers have a greater fitness than non- carriers and homozygote carriers.

Insertions and Deletions. The insertion or deletion of a base pair within a gene can have disastrous effects. Because the messenger RNA (mRNA) reads a sequence of three base pairs at a time as a codon, the deletion or insertion of a base pair will cause a frame shift that causes nonsense readings downstream of the insertion or deletion and will result in the synthesis of a nonfunctional protein.

Chromosomal Mutations. Chromosomal mutations are major changes in the structure of the genome. An example is the mutation that separates humans and chimpanzees: Chimpanzees and gorillas have 24 pairs of chromosomes, whereas humans have only 23 pairs. Sometime after the hominid-chimpanzee speciation about 6.5 million years ago, two of the chimpanzee's chromosomes fused together to become one chromosome that became the human chromosome 2.

As pointed out above, most mutations are deleterious. Sometimes, whole genes are deleted. Williams syndrome is a rare neurological disorder that causes extreme sociality and learning disorders in children. It results from the deletion of 28 genes on at the 11th band on the long arm of the seventh chromosome.[8,9,10]

A common chromosomal mutation is the duplication of an entire gene, which can have one of two consequences: (1) the duplicate gene can continue to function and produce its intended products, so that those products are produced at an increased rate, or (2) the original gene can continue its original function and the new gene can mutate and assume a new function. An example is the evolution of the antifreeze glycoproteins (AFGP) found in Arctic and Antarctic fish that prevent the freezing of the blood of these fish. The AFGP gene is structurally very similar to the pancreatic trysinogen gene and is believed to have evolved as a duplication of the pancreatic trysinogen gene.[11] This mutation occurred independently in both the Arctic and Antarctic fish. Such an independent evolutionary process is called convergent evolution. In humans, who carry about 21,000 genes, gene duplication occurs about every 5,000 years, so that in a population of 7 billion humans, 1.4 million individuals should expect a gene duplication in their genome, but most of these mutations will be nonfunctional, and a deleterious mutation may cause a non-viable fetus that would not be born.

Stretches of non-coding DNA frequently contain repeated base pairs and are vulnerable to being duplicated during copying, a process called slippage, which causes the new strand to be out of sync with the copied strand so that a section of the copied strand is missed and not copied into the new strand.

Another error is the transposition of a gene, in which a gene can be transposed to a new location on the chromosome or even moved to a different chromosome. Such mutations are called jumping genes. If a transposable element inserts itself into another gene, the element will corrupt the gene and make it nonfunctional. Transposition usually changes the length of a chromosome, so that when the chromosome attempts to join with its homologous counterpart during sexual reproduction, there will be a mismatch and recombination between the two chromosomes will be unsuccessful. A deletion of part of a chromosome can have the same effect. Such mutations usually result in miscarriages in sexually reproducing species.

Finally, whole sections of a chromosome may be transferred to a different location on the same chromosome or inserted into another chromosome, or it may be inverted, or two chromosomes may fuse together, as happened in the human chromosome after the hominin-chimpanzee speciation. Such mutations are usually deleterious and usually result in nonviable fetuses in sexual reproduction.

Pleiotropy and Punctuated Equilibrium

Early paleontologists were puzzled by the fact that the fossil record did not show gradual changes in fossils as they might have evolved from one species to another, but it appeared that new species appeared abruptly between layers in the geologic record that were deposited over time. These facts were a challenge to Darwin's hypothesis of gradual change, or gradualism, in the evolutionary process. In 1972, Niles Eldredge and Stephen Jay Gould published a paper in which they proposed their theory of punctuated equilibrium, in which new species emerge rapidly after a long period of stasis between speciation.[12] Speciation usually occurs in small populations that are separated from a larger one and exposed to a different environment.[13]

One mechanism for explaining punctuated equilibrium is pleiotropy. The expression of most coding genes is controlled by other genes called regulatory or master genes. Because regulatory genes control the expression of more than one coding gene, they can control more than one trait or phenotype, a process called pleiotropy. A mutation in a pleiotropic regulatory gene can significantly effect the morphology of a species and can indeed lead to the creation of a new species. Although the human and chimpanzee genomes are strikingly similar and differ by less than one percent in their coding regions, the morphological and behavioral differences between the two species are vast. The major genetic differences seem to be when and the rate at which those genes are expressed. This led Mary-Claire King and Allan Wilson to speculate that major genetic differences are mutations in the regulatory genes of the two species. Pleiotropic mutations can explain the rapid evolution of a new species.[14]

End Notes

Chapter 1

1. George Santayana, *Tipperary Soliloquies in England*, London, Scribner 1924: 102.
2. Chris Hedges, *What Every Person Should Know about War,* New York, Free Press 2003: 1.
3. N.P. Gleditsch, P. Wallenstein and M. Eriksson, "Armed Conflict 1946–2001: A New Data Base," *Jor. Peace Research* 39 (2002): 615–637.
4. Adam Crowell, "War Deaths Said to Top 100,000," *New York Times*, June 26, 2013.
5. Lewis F. Richardson, *Statistics of Deadly Quarrel,* Pittsburg 1969, Boxwood Press: 6.
6. Steven A. LeBlanc, Battles: *The Myth of the peaceful, Noble Savage,* New York 2003. Saint Martins Press: 56-57.
7. Margret Mead, "Alternatives to War," in *War,* ed. M. Fried, M. Harris and R. Murphy, Garden City NJ, Natural History Press: 215-228.
8. Immanuel Kant, *Perpetual Peace (1795)*, New York, Liberal Arts Press, 1957.
9. www.ssa.gov/history/acharter2.htm/ (accessed June 12, 2003).
10. John Keegan, *A History of Warfare*, New York 1993, Knopf: 32-34, 59-61, 73.
11. Gerhard von Glahn, *Law Among Nations: An Introduction to Public International Law,* New York 1965, Macmillan Co.: 461-462.
12. Charles S. Rhyne, *International Law: The Substances, Procedures and Institutions of World Peace with Justice,* Washington 1971, CLB Publishers: 265-272.
13. Herbert Spencer, *Social Statics, or the Conditions Essential to Human Happiness*, London 1851, Williams and Norgate.

14. J.M.G. van der Dennen, *The Origin of War* 2002, http;//rint.rechten.rug.nnl/rth/dennen/. (accessed May 17, 2010).

Chapter 2

1. Jane Goodall, *The Chimpanzees of Gombe: Patterns of Behavior,* Cambridge, MA, Harvard Univ. Press 1986: 534.
2. Speciation dates in Figure 2.1 are from Table 7 of N. Patterson, D.J. Richter, S. Gnerre, E.S. Lander, and D. Reich, "Genetic Evidence of Complex Speciation of Humans and Chimpanzees," *Nature* 441 (2006): 1103–1108, calibrated for a human-chimpanzee speciation date of 6.5 million years ago. This date was obtained by averaging the earliest date of *Orrorin tugenensis*, believed to be the oldest hominin. (Yoshihiro Sawada, Martin Pickford, Brigite Senul, Tetsumaro Itaya, Masayuki Hyodo, et al., "The Age of *Orrorin tugenensis*, an Early Hominind from the Yugen Hills, Kenya," *Comptes Rendus Palevol.* 1 (2002): 292-303; Michel Brunet, Franck Guy, David Pilbeam, Hassane Taisso Mackaye, Andossa Liklus, et al., "A New Hominid from the Upper Miocene of Chad, Central Africa," *Nature* 318 (2002): 145–151).
3. A.J. Harcourt, "The Social Relations and Group Structures of the Wild Mountain Gorilla," in *Great Apes,* ed. David Hamburg and Elizabeth R. McCown, Menlo Park, CA, Benjamin/Cummings Publishing Co. 1979: 187–192.
4. E.J. Stokes, "Within-Group Social Relationships among Females and Adult Males in Wild Western Lowland Gorillas (*Gorilla gorilla gorilla*)," *Am. Jor. Primatology* 64 (2004): 233–246.
5. B.M.F. Galdikas, "Orangutan Adaptation at Tanjung Putting Reserve: Mating and Ecology," in *The Great Apes,* ed. David A. Hamburg and Elizabeth R. 0McGowan, Menlo Park, CA, Benjamin/Cummings 1979: 195– 234.
6. Richard Wrangham, "Out of the Pan, Into the Fire: How Our Ancestors' Evolution Depended on What They Ate," in *Tree of Origins What Primate Behavior Can Tell Us about Human Social Evolution*, ed. Franz de Waal, Cambridge, MA, Harvard University Press 2001: 119–144.
7. Ronald M. Nowak, *Walker's Primates of the World*, Baltimore,

MD, Johns Hopkins Univ. Press 1999: 178–182.

8. M. Bermejo, "Home-Range Use and Intergroup Encounters in Western Gorillas (*Gorilla g. gorilla*) at Lossi Forest, North Congo," *Am. Jor. Primatology* 64 (2004): 223–232.
9. Christopher Bohem, "Ancestral Hierarchy and Conflict," *Science* 336 (2011): 844–847.
10. Charles Darwin, *The Descent of Man, and Selection in Relation to Sex*, London, John Murray, 1871, facsimile reproduction, Princeton, Princeton Univ. Press 1981: vol. 1, 163-164.
11. Sally McBrearty and Nina G. Jablonski, "First Fossil Chimpanzee," N*ature* 437 (2005): 105–108.
12. Nowak 1999: 183.
13. George P. Murdock, *Ethnographic Atlas*, Pittsburgh, PA, Univ. of Pittsburgh Press 1967.
14. Goodall 126-133.
15. Ibid: 128.
16. Ibid: 412.
17. Ibid: 342–343.
18. Ibid: 489.
19. Ibid: 493.
20. Richard Wrangham, "Evolution of Coalitionary Killing," Yearbook of Physical Anthropology 1999: 1-30.
21. After Goodall 1986: 229.
22. Goodall 1986: 549–559.
23. Wrangham 1999,. Table 4.
24. Goodall 1986: 83
25. Ibid: 503-506.
26. Ibid: 514.-518
27. Ann E. Pusey, "Of Genes and Apes: Chimpanzee Social Organization and Reproduction," in *Tree of Origin: What Primate Behavior Can Tell Us about Human Social Evolution,* ed. F.B.M. de Waal, Cambridge MA, Harvard Univ. Press 2001: 9–38.
28. Christopher Boesch and Hedwige Boesch-Achermann, *The Chimpanzees of the Taï Forest*, Oxford UK, Oxford Univ. Press 2000: 135–138.
29. Ibid: 142–145.
30. Jennifer M. Williams, W. Oehlert, John V. Carlis and Anne E.

Pusey, "Why Do Male Chimpanzees Defend a Group Range," *Animal Behavior* 68 (2004): 528–532..
31. Richard Wrangham, "Why Apes and Humans Kill," in *Conflict*, ed. Martin Jones and A.C. Fabian, Cambridge UK, Cambridge Univ. Press 2006.
32. Nowak 1999: 185.
33. Ibid.
34. Franz de Waal and Franz Lanting, *Bonobo: The Forgotten Ape*, Berkeley, Univ. of California Press 1997: 113.
35. Nowak 1999: 185.
36. de Waal and Lanting 1997: 113.
37. Ibid: 106-1074.
38. Ibid: 109-110.
39. Ibid.
40. Ibid: 111-112.
41. Ibid: 113.
42. Amy Randall Parish, "Female Relationships in Bonobos (*Pan Panscus*)," *Human Nature* 7 (1986): 262.
43. Franz de Waal, *Our Inner Ape*, , New York, Riverhead Books 2008: 109.

Chapter 3

1. John Keegan, *A History of Warfare*, New York, Knopf 1993: xvi.
2. Lawrence H. Keeley, *War before Civilization: The Myth of the Peaceful Savage*, Oxford, UK Oxford Univ. Press 1996: 17–24.
3 . Mark Howard Ross, "Political Decision Making and Conflict: Additional Cross-Cultural Codes and Scales," *Ethnology* 22 (1983): 169–192.
4. Keith Otterbein, *The Evolution of War: A Cross-Cultural Study* 3rd ed., New Haven, CT, AF Press 1989: 143–144, 148.
5. Joseph J. Jorgensen, *Western Indians: Comparative Environments, Languages and Cultures of 172 Western American Indian Tribes,* San Francisco, CA, W.H. Freeman and Co. 1980: 503–6, 509–16, 613–614.
6. Marilyn Keyes Roper, "A Survey of the Evidence for Intrahuman Killing in the Pleistocene," *Current Anthropology* 10 (1969): 427–

459.

7. Johannes Maringer and Hans-Georg Bandi, *Art in the Ice Age: Spanish Levant Art: Arctic Art*, London, George Allen and Unwin 1953: plate 173.
8. Dawn by Schelay Richardson, after Johannes Maringer and Hans Georg Bandi, *Art in the Ice Age: Spanish Levant Art: Arctic Art*, London, George Allen and Unwin 1953: plate 158.
9. J.L. Arsyuaga, *The Neanderthal's Necklace: In Search of the First Thinkers*, New York, Four Walls, Eight Windows 1969: 58.
10. Keeley 1996: 37.
11. Ibid.
12. Arthur Ferrill, *The Origins of War*, London, Thames and Hudson 1985: 23-24.
13. Otterbein 1989: 63–70; 148–149. (If column 23, 24, or 25 contained a 3, indicating "land" as a reason for going to war, the society was counted as territorial.).
14. Elizabeth Cashdan, "Territoriality among Human Foragers: Ecological Models and an Application to Four Bushman Groups," *Current Anthropology* 24 (1983): 47–66.
15. Rada Dyson-Hudson and Eric A. Smith, "Human Territoriality: An Ecological Reassessment," *Am. Anthropologist* 80 (1978): 21–41.
16. Edward O. Wilson, *On Human Nature*, Cambridge MA, Harvard Univ. Press 1978: 108-109.
17. Ibid.
18. Ibid.
19. Steven A. LeBlanc, *Constant Battles: The Myth of the Peaceful, Noble Savage*, New York, St. Martin's Press 2003: 122.
20. J. Morgan, *The Life and Adventures of William Buckley: Thirty-two Years a Wanderer amongst the Aborigines*, Canberra, Australian National University Press 1852, quoted in ibid: 123.
21. Ibid, quoted in LeBlanc 2003: 114.
22. W. Lloyd Warner, "Murngin Warfare," *Oceania* 1 (1930–1): 457–494.
23. LeBlanc 2003: 123.
24. Frank Raymond Secoy, *Changing Military Patterns of the Great Plains Indians*, Lincoln, NE, Univ. of Nebraska Press 1993: 6f.
25. John D. McDermott, *A Guide to the Indian Wars of the West*, Lincoln, NE, Univ. of Nebraska Press 1998: 46.

26. Thomas E. Mails, *The Mystic Warriors of the Plains: The Culture, Arts, Crafts of the Plains Indians,* New York, Marlowe and Company 1972: 516-518.
27. Ibid: 554–558.
28. Ibid: 570.
29. Ibid: 572–575.
30. Patrick M. Malone, *The Sulking Way of War: Technology and Tactics among the New England Indians*, Baltimore, MD, Johns Hopkins Univ. Press 1991: 16.
31. Ernest S. Burch Jr., "Eskimo Warfare in Northwest Alaska," *Anthropological Papers of the University of Alaska* 16 (1974): 1–14.
32. Ibid: 10-11.
33. Ibid: 3.
34. LeBlanc 2003: 118.
35. Edward W. Nelson, "The Eskimo about Bering Strait," *Annual Report 18, Bureau of American Ethnology 1897, part 1*, Washington: 327.
36. Burch 1974: 8.
37. Ibid: 10-11.
38. Napoleon A. Chagnon, *Yąnomamö* 5th ed., Belmont, CA, Wadsworth Thomson Learning 1997: 60-69.
39. Ibid.: 160.
40. Ibid.
41. Ibid: 185.
42. Ibid: 178–180.
43. Ibid: 180-181.
44. Ibid: 186–189.
45. Ibid: 190.
46. Ibid: 198-199.
47. Ibid: 189-190.
48. Leopold Pospisil, *The Kapauku Papuans of West New Guinea*, New York, Holt, Rinehart and Winston 1963: 42–44.
49. Jared Diamond, *Collapse: How Societies Chose to Fail or Succeed,* New York, Viking 2005: 280.
50. Ibid: 45.
51. Ibid, 58; Jared Diamond, "Vengeance Is Ours," *The New Yorker*,

21 April 2008: 74–87.
52. Pospisil 1963: 59.
53. from the Vinkhuijzen Collection of Military Costume Illustration, New York Public Library Digital Collection.
54. Ibid: 60-61.
55. Diamond 2005: 57.
56. Ferrill 1985: 99–106.
57. Ibid.
58. Thomas Gregor, "Uneasy Peace: Intertribal Relations in Brazil's Upper Xingu," in *The Anthology of War*, ed. Jonathan Hass, Cambridge UK, Cambridge Univ. Press 1990: 110.
59. Ibid: 114.
60. Ibid.: 117.
61. Robert Knox Dentan, *The Semai: Nonviolent People of Malaya*, New York, Holt, Rinehart and Winston 1968: 55.
62. Ibid.: 58-59.
63. Donald F. Fry and Patrick Söderberg, "Lethal Aggression in Mobile Forager Bands and implications for the Origins of War," *Science* 341 (July 19, 2003): 270-273.
64. Otterbein 1989: 63–70.
65. Jorgensen 1980.
66. Keeley 1996: 200.
67. Chagnon 1997: 205.
68. Mails 1972: 12.
69. Charles Darwin, *The Descent of Man, and Selection in Relation to Sex,* London, John Murray, 1871, facsimile reproduction, Princeton, Princeton Univ. Press 1981: vol. 1, 166.
70. Jean-Jacques Rousseau, "Discourse on the Origin and Basis of Inequality," in *The Essential Rousseau,* trans. Lowell Bair, New York, Penguin Books 1983: 184.
71. Ibid: 144.
72. David Fabbro, "Peaceful Societies: An Introduction" *Jor. of Peace Research* 15 (1978): 67–83.
73. Keeley 1996: 37.
74. Ibid.
75. Kelly 2000: 52.
76. Ibid: 133.
77. Ibid.: 159.

78. Otterbein 2004: 219-220.

Chapter 4

1. Herbert Spencer, *Social Statics, or the Conditions Essential to Human Happiness*, London, Williams and Norgate 1851: 455.
2. Michael Balter, "Seeking Agriculture's Ancient Roots," *Science* 216 (2007): 1830–1835.
   3. George Wilcox, "The Roots of Cultivation in Southwestern Asia," *Science* 341 (2013): 39–40.
4. S. Lev-Yadun, A. Gopher, and S. Abbo, "The Cradle of Agriculture,"*Science* 88 (2000): 1602–1603.
5. Jared Diamond, *Guns, Germs and Steel,* New York, W.W. Norton & Co. 1999: 100.
6. Brenda Fowler, "African Pastoral: Archaeologists Rewrite History of Farming," *New York Times*, July 27, 2004.
7. Balter 2007.
8. Wilcox 2013.
9. Diamond 1999: 106.
10. John-Pierre Boquet-Appel, "When the World's Population Took Off: The Springboard of the Neolithic Demographic Transition," *Science* 333 (2011): 560–561.
11. Diamond 1999: 273–278.
12. Drawn by Schelay Richardson. From *The Origins of War* by Arthur Ferrill, London, Tames and Hudson, 1985: 29.
13. After James Mellaart, From *The Origins of War* by Arthur Ferrill, London, Tames and Hudson, 1985: 31.
14. Arthur Ferrill, *The Origins of War*, London, Tames and Hudson: 28-30.
15. .Thorkild Jacobson, "Sumer," in *The Penguin Encyclopedia of Ancient Civilizations,* ed. Arthur, Cotterell, New York, Penguin Books 1980: 72-84.
16. Ibid: 76.
17. Robert L. Carneiro, "A Theory of the Origin of the State," *Science* 169 (1970): 733–738.
18. Jacobson 1980: 76.
19. Ibid: 75.

20. Ibid: 78-79.
21. Samuel Noah Kramer, *The Sumerians: Their History, Culture and Character,* Chicago, IL, Univ. of Chicago Press 1973: 323.
22. Napoleon A. Chagnon, "The Culture-Ecology of Shifting (Pioneering) Cultivation among the Yąnomamö Indians," *Proceedings, VIIIth International Congress on Anthropological and Ethnological Sciences* 3 (1968): 249, 251.
23. Courtesy of the Metropolitan Museum, New York.
24. Colin Walters, "Ancient Egypt," in *The Pengui Encyclopedia of Ancient Civilizations*, ed. Arthur Cotterell, New York, Penguin Books 1980: 22–46.
25. Ibid.
26. Jane McIntosh and Clint Twist, *Civilizations: Ten Thousand Years of Ancient History*, New York, DK Publishing 2001: 88.
27. Ibid: 89.
28. Ibid: 73.
29. Ted Goebel, Michael R. Waters, and Dennis H. O'Rourke, The Late Pleistocene Dispersal of Modern Humans in America," *Science* 319 (2008): 1497–1502.
30. McIntosh and Twist, 2001: 184.
31. Ibid: 221–222.
32. Arthur Ferrill, *Origins of War*, London, Thames and Hudson 1985: 88.

Chapter 5

1. Charles Darwin, *The Descent of Man, and Selection in Relation to Sex,* London, John Murray, 1871, facsimile reproduction, Princeton,Princeton Univ. Press 1981: vol. 1, 106.
2. Tom Segev, *One Palestine, Complete*, New York, Metropolitan Books 2000: 9.
3. Alissa Rubin, "As Afghan Forces Kill, Trust Is Also a Causality," *New York Times,* October 22, 2012.
4. Global Security, "The World at War," globalsecurity.org/military/world/war/index.html, (accessed September 13, 2014).
5. Correlates of War Project, Inter-State War and Intra-State War Data Sets

www.correlatesofwar.org/cow2%20Data/WarData_New/Inter-stateWarData_v4.0.csv, and www.correlatesofwar.org/cow2%20Data/WarData_New/Intra-stateWarData_v4.1.csv (Accessed August 24, 2014)

6. Wikipedia, Lists of Wars: 1945-1989; 1990-2002; 2003-2010; 2011-present, Wikipedia/wiki, List_of_Wars . . .
7. D.A. Summerfield, *Global Trends, Alternative Worlds*, Washington, DC,
8. Exodus 15:3–6, *New English Bible* (NEB), New York, Oxford University Press 1970.
9. Deuteronomy 20:1–17, NEB.
10. Karen Armstrong, *Islam: A Short History*. New York, Modern Library 2000: 30.
1J. Choi and S. Bowles, "The Coevolution of Parochial Altruism and War," *Science* 318 (2007): 636–640.
12. Darwin 1871: 93.
13. Daniel Patrick Moynihan, *Pandemonium:Ethnicity in International Relations*, New York, Oxford Univ. Press 1993: 30f.
14. Chas W. Freeman, Jr., *The Diplomat's Dictionary*, Washington DC, National Defense University Press 1994: 231.
15. John King Fairbanks, *China: A New History*, Cambridge MA, Harvard Univ. Press 1992: 147.
16. Gerald Prunier, *The Rwanda Crisis: History of a Genocide*, New York, Columbia Univ Press 1995: 13–14, n. 31.
17. Edward O. Wilson, *Sociobiology: The New Synthesis*, Cambridge, MA., Harvard Univ. Press 2000: 249.
18. Ibid: 249, 286.
19. Charles Darwin, *The Correspondence of Charles Darwin, 1821–65*, vols. 1–12, ed. Fredrick H. Burkhardt et al., Cambridge UK, Cambridge Univ. Press 1983–2001: v. 7, p. 81.
20. Jared Diamond, *The Third Chimpanzee: The Evolution and Future of the Human Animal,* New York, HarperCollins 1992: 220.
21. R. Paul Shaw and Yuwa Wang, *Genetic Seeds of Warfare: Evolution, Nationalism and Patriotism*, Boston, MA, Urwin Hyman 1989: 80.
22. Editorial, "Stateless in the Dominican Republic," *The New York Times*, July 11, 2015

23. Ellen Berry, "Russian Governor Signs up Cossacks to Police Migrants," *New York Times*, August 4, 2012.
24. Edward Wong, "China: Muslim Fasting Discouraged," *New York Times*, August 4, 2012.
25. Alissa J. Rubin and Sanngar Rahimi, "Afghan Officials Cite Revenge Killings in Latest Outbreak of Ethnic Hatred," *New York Times*, August 4, 2012.
26. Jeffery Gettleman, "Riots Battle Kenya as Rivals Declare Victory," *New York Times*, December 30, 2007.
27. Quran: 9.5
28. Quran: 9.25
29. R.N. Dandekar, "Then Bhagavad-Gita," in *Sources of Indian Tradition*, vol. I, *From the Beginning to 1800*, 2nd ed. ed. Ainslie Embree, New York, Columbia Univ. Press 1988:280.
30. John King Fairbanks, *China: A New History*, Cambridge MA., Harvard Univ. Press: 108-112.
31. Michael Walzer, *Arguing About War*, New Haven, Yale Univ. Press 1
32. Shaw and Wang 19890:143
33. Xiaojing Xu, Xiangyu Zuo, Xiaoying Wang and Shihui Han, "Do You Feel My Pain? Racial Group Membership Modulates Emphatic Neural Response," *Jor. of Neuroscience,* 29 (2009) 8525-8529
34. Lucy R. Osborne, Martin Li, Barbra Pober, David Chitayat, Joannn Boduirtha, et al., "A 10 Million-Base Pair Inversion Polymorphism in Families with Williams-Beuren Syndrome," *Nature Genetics* 28 (2001): 321–325.
35. Ursula Bellugi, Loz Lichtenberger, Debra Mills, Albert Galaburda, and Julie. Korenberg, "Bridging Cognition: The Brain and Molecular Genetics from Williams Syndrome," *Trends in Neuroscience* 22 (1999): 197–207.
36. Andreas Meyer-Lindenberg, Carolyn B. Mervis, and Karen Faith Berman, "Neural Mechanisms in Williams Syndrome: A Unique Window to Genetic Influences on Cognition and Behavior," *Nature Reviews Neuroscience,* 7 (2006): 380–393.

Chapter 6

1. A. Shadid, "At Night, Protest Gives Way to Poetry," *The New York Times,*February 7, 2011.
2. Isabel KOershner, "Israel Sees Escalation in Gazans' Longer-Range Strikes," New *York Times*, February 29, 2007.
3. Martin Gilbert, *The First World War: A Complete History,* New York, Henry Hold and Co. 1994: 72.
4. John Keegan, *The First World War,* New York, Alfred Knopf 1999: 295, 298-299.
5. Ibid.
6. John Keegan and Richard Holmes, *Soldiers: A History of Men in Battles*, New York, Viking 1985: 49.
7. Christopher Tyerman, *Fighting for Christendom: Holy War and the Crusades,* New York, Oxford Univ. Press 2004: 3-4.
8. Ben Hubbard, "In Jordon Town, Syria War Inspires Jihadist Dreams," *New York Times,* April 12, 2014.
9. Ceylan Yeginsu, "ISIS Draws a Steady Stream of Recruits From Turkey,"*New York Times*, September 15, 2014.
10. Henry de Mann, *The Remaking of a Mind: A Soldier's Thoughts on War and Reconstruction,* New York, C. Scribner's Sons 1920: 198-199.
11. D.M. Cook, *Spitfire Pilot*, London, Faber and Faber 1942: 28, 30f.
12. Roderick Chisholm, *Cover of Darkness*, London, Chatto and Windus 1973: 71.
13. H.F. Pringle, *Theodore Roosevelt: A Life,* New York, Harcourt Brace 1956: 181.
14. Robert E. Lee, comments to General James Longstreet on seeing the federal charge repulsed during the Battle of Fredericksburg, February 13, 1862 in Edward Porter Alexander, *Military Memoirs of the Confederacy: a Critical Narrative*, New York, C. Scribner's Sons 1907: 302.
15. The HBO miniseries *Band of Brothers* presents a comprehensive and realistic view of the history, training, and cohesion of Easy Company of the US Army Airborne Paratroop Division during World War II.
16. Donavan Campbell, *Joker One: A Marine Platoon's Story of Courage, Leadership, and Brotherhood,* New York, Random House 2009: 8.

17. Keegan and Holmes 1985: 19.
18. Gwynne Dyer, *War*, Toronto, Stoddard 1985: 127.
19. Keegan and Holmes 1985: 53.
20. William Manchester, *Goodbye Darkness: A Memoir of the Pacific War,* Boston, Little Brown & Co. 1980: 451-452.
21. Dave Grossman, *On Killing: The Psychological Cost of Learning to Kill in War and Society,* Boston, Little Brown and Co. 1995: 3f.
22. Quoted in Ibid: 11.
23. Robin Tolmach Lakoff, "From Ancient Greece to Iraq, the Power of Words in Wartime," *Science Times, The New York Times*, May 18, 2004.
24. David A. Graff, *Medieval Chinese Warfare 300–900*, London, Routledge 2002: 20.
25. Nichols D. Kristoff, "The Face of Genocide," *The New York Times*, November 19, 2006.
26. H. Strachan, *The First World War,* vol. 1,. *To Arms.* Oxford UK, Oxford Univ. Press 2001: 160–162.
27. Keegan 1999: 72.
28. John Muller, "The Iraq Syndrome," *Foreign Affairs* 84 (Nov./Dec. 2005): 44– 54.
29. Robert A. Pape, *Dying to Win: The Strategic Logic of Suicide Terrorism,* New York, Random House 2005: 13
30. Ibid: 139.
31. Israeli Ministry of Foreign Affairs, "Suicide and Other Bombing Attacks on Israel since the Declaration of Principles, September 1993," 2007, www.mfa.gov.il/MFA/Terrorism+Obsticle+to+peace/Palestinian+terror+since+2000/suicide+and+other+b ombing+attacks+in+Israel+since.htm (accessed July 18, 2014).
32. Pape 2005: 132.
33. Ibid.
34. Nasra Hassan, "An Arsenal of Believers," *The New Yorker*, November 19, 2001,
35. Ibid.
36. Ibid.
37. Andrea Elliott, "Where Boys Grow up to Be Jihadists," *New York Times*, November 25, 2007.
38. Amnesty International, "Without Distinction: Attacks on

Civilians by Palestinian Armed Groups," July 2002.
39. Pape 2005: 172-173.
40. Hassan 2001.
41. Ibid.
42. Ibid.
43. Ibid.
44. Ibid.
45. Ibid.

Chapter 7

1. Edward O. Wilson, *The Social Conquest of the Earth*, New York: W.W. Norton 2012: 17.
2. quoted in Mary Ann Glendon, *A World Made New: Eleanor Roosevelt and the Universal Declaration of Human Rights*, New York, Random House 20`:42.
3. Bert Hölldobler and Edward O. Wilson, *The Superorganism*, New York, W.W. Norton 2009: 9.
4. Edward O. Wilson, *Sociobiology: The New Synthesis*, Cambridge, MA., Harvard Univ. Press 2000: 419–435.
5. Dorothy L. Cheney and Robert M. Seyfarth, *Baboon Metaphysics*, Chicago, IL, Univ. of Chicago Press 2007: 10.
6. Ronald M. Nowak, *Walker's Primates of the World*, Baltimore, MD, Johns Hopkins Univ. Press 1999: 148.
7. William Norton Wheeler, *The Social Insects: Their Origin and Evolution.* London, Kegan Paul, Trench, Trubner (1928). reprint, New Delhi, Discovery Publishing House 2006: 230.
8. William Norton Wheeler, "The Ant Colony as an Organism," *Jor. of Morphology* 22 (1911): 307–325.
9. Hölldobler and Wilson, 2009: 8–12.
10. Ibid: 8–10.
11. Herbert Spencer, *Principles of Sociology*. New York, D. Appleton & Co. 1881: 467f.
12. Robert L. Carneiro, "The Chiefdom Precursor of the State," in *The Transition to Statehood in the New World*, ed. G.D. Jones and R.R.Kautz, Cambridge, UK, Cambridge Univ. Press 1981: 37–79.
13. Jared Diamond, *Guns, Germs and Steel,* New York, W.W orton & Co. 1999:: 273.
14. Walter Lippmann, "An Image of Man for a Liberal Democracy,"

in *The Essential Lippmann,* ed. Clinton Rossiter and James Lare, New York, Random House 1963: 162–168.

15. Herbert Spencer, "The Social Organism," *The Westminster Review,* January 1860. Reprinted in Herbert Spencer, *Essays: Scientific, Political, and Speculative,* London, Williams and Norgate 1863: 143–185.
16. Herbert Spencer, *Principles of Sociology* 3rd ed., New York, D. Appleton & Co. 1896: 595.
17. Much of this section is drawn from Martin A. Nowak, Corina E. Tarnita, and Edward O. Wilson, "The Evolution of Eusociality," *Nature* 466 (2010): 1057–1062.
18. Darwin 1871: 89.
19. Ibid: 163.
20. Ibid: 162–166.
21. Ibid: 163.
22. Kenichi Aoki, "A Condition for Group Selection to Prevail Over Counteracting Individual Selection," *Evolution* 36 (1982): 832-842.
23. Samuel Bowles, "Group Competition, Reproductive Leveling, and the Evolution of Human Altruism," *Science* 314 (2006): 1569-1572..
24. Charles Goodnight and Lori Stevenson, "Experimental Studies in Group Selection: What Do They Tel Us About Group Selection in Nature," *American Naturalist* 150 (1997): 559-579.
25. Anne Traulsen and Martin A. Nowak, "Evolution of cooperation by multilevel selection," *Proc. National Acd. of Science*, 103 (2006): 10952-10955.
26. Peter Turchin, "Warfare and the Evolution of Social Complexity: A Multilevel Selection Aproach," hppt://Peter.Turchin@ uconn.edu.(accessed September 1, 2014)
27. Michael J. Wade, "Soft Selection, Hard Selection, Kin Selection and Group Selection," *The American Naturalist,* 125 (1985): 61-73.
28. David Sloan Wilson, and Elliott Sober, "A Theory of Group Selection," Proc. *National Acd. of Science*, 72 (1975): 143-146..
29. Emanuel Kant, "Beantwortung der Fraga: Was ist Aufklärung?" *Berlinische Monatsschrift*, December 1784, https://web.cn.edu/kwheeler/documents/What_is_Enlightenment.

pdf (accessed July 20, 2014).
30. John Locke, *An Essay Concerning Human Understanding*, Menston, England, Scholar Press 1970: 438.
31. .Wilson 2012: 58.

Chapter 8

1. Edwin O. Wilson, *Sociobiology* (25th Anniversary edition), Cambridge MA, Harvard University Press 2000: 248.
2. Steven C. Maxon and Andrew Canastar, "Genetic Aspects of Aggression in Nonhuman Animals," in *Biology of Aggression*, ed. Randy J. Nelson, New York, Oxford Univ. Press 2006: 3–19.
3. Larry J. Silver, "Neurobiology of Aggression and Violence," *Am. Jor. of Psychiatry* 156 (2008): 429–442.
4. Stephen C. Gammie and Joseph Lonstein, "Maternal Aggression" in *Biology of Aggression*, ed. Randy J. Nelson, Oxford, UK, Oxford Univ. Press 2006: 250–274.
5. Robert Plomin, John C. Defries, Gerald E. Mclearn, and Michael Rutter, *Behavioral Genetics* 3rd ed., New York, W.H. Freeman 1997: 63.
6. Kristi M.J. Lagerspetz and Kari Y.H. Lagerspetz, "Genes and Aggression," in *Aggressive Behavior: Genetic and Neural Approaches*, ed. Edward C. Simmel, Martin E. Hahn, and James K. Walters, Hillsdale, NJ, Euribaum Associates 1983: 89–101.
7. Kristi M.J. Lagerspetz and Kari Y.H. Lagerspetz, "The Expression of the Genes of Aggression in Mice: The Effect of Androgen on Aggression and Social Behavior in Females," *Aggressive Behavio,* 1 (1983): 291-296.
8. Kristi Lagerspetz, "Genetic and Social Causes of Aggressive Behavior in Mice," *Scandinavian Jor. of Psychology* 2 (1961): 167–173.
9. Kristi Lagerspetz, *Studies on the Aggressive Behavior of Mice,* Helsinki, Finland, Suomalainen Tiereakatemia 1964.
10. Kari Y.H. Lagerspetz, Rauno Tirri, and Kristi M.J. Lagerspetz, N urochemical and Endocrinological Studies of Mice Selectively Bred for Aggression," *Scandinavian Jor. of Psychology* 9 (1968): 157–160.

11. Kristi Lagerspetz, "Genetic and Social Causes of Aggressive Behavior in Mice,' Scandinavian Jor. of Psychology 2 (1961) 167-173
12. Lagerspetz 1964
13.. Data from Joshua Goldstein, *War and Gender*, Cambridge UK, Cambridge Univ. Press 2001: 145.
14. J. Ghodsain-Carpey and L.A. Baker, "Genes and environmental influences on aggression in 4- to 7-year old twins," *Aggressive Behavior*, 13 (1987):173-186.
15. T.C. Edelbrock, R. Rende, R. Plomin and L.A. Thompson, "A twin study of competence and problem behavior in childhood and early adolescence," *Jor. Child Psychology and Psychiatry,* 36 (1995): 775-785.
16. T.C. Ely, P. Lichteenstein and J. Stevens, "Sex differences in the etiology of aggressive and nonaggressive and antisocial behavior: Result from two twin studies," *Child Development*, 70 (1999): 155-168.
17.. R.M. Boyar, R.S. Rosenfeld, S. Kapen,. "Human Puberty Simultaneous Augment Secretion of Lutenizing Hormone and Testosterone During Sleep," *Jor. of Clinical Investigation;* 6 (1991): 609-618.
18. Kerrin H. Christansen, "Serum and Saliva Sex Hormone Levels in !Kung San Men," *Am. Jor. Physical Anthropology*, 86 (1991): 37-44.
19. Ibid.
20. Private communications with Mark Breedlove and Randy Nelson.
21. Robert M. Sapolsky, *The Trouble with Testosterone: Essays on the Biology of the Human Predicament*, New York, Simon and Schuster 1997: 159.
22. Goldstein 2001: 150.
23. Allan Mazur and Theodore A. Lamb, "Testosterone , Status, and Mode in Human Males," *Hormones and Behavior* 14 (1980): 236-246.
24. P.J. Perry, K.H. Anderson and S.R. Yeats, "Illicit anabolic use in athletes: A case series analysis," *Am. Jor. of Sports Medicine* 18 (1990): 422-429.
25. W.R. Yeats, P. Perry & and S. Murphy, "Aggression and hostility in anabolic steroid users," *Biological Psychiatry* 31 (1992): 1232-

1244
26. W.R. Yeats, P.J. Perry & K.H. Anderson, "Illicit anabolic steriod use: a controlled personality study," *Acta Psychiaric Scandinavia* 81 (1997) 5548-5550.
27.. Mazur and Lamb 1980.
28. Ibid.
29..E. Gonzales-Bono, A. Salvador, M.A. Serrano and J. Ricarte, "Testosterone, Cortisol and Mood in a Sports Team Competition," *Hormones and Behavior* 35 (1999): 55-62.
30. M.A. Serrano, A. Salvador, E. Gonzalez-Bono, C Sanchis and E. Suay, "Hormonal Response to Competition," *Psycothema* 12 (2000): 440-444.
31. Allen Mazur, Alan Booth and James M. Dabbs, "Testosterone and Chess Competition," *Social Psychology Quarterly* 55 (1992): 70-77.
32. Mazur and Lamb 1980.
33. James M. Dobbs, Denise de LaRue and Paula M. Williams, "Testosterone, Occupational Choice: Actors, Ministers and Other Men," *Jor. Personality and Social Psychology*, 59 (1990): 1261-1265.
34. Rae Silver, "Environmental Factors Influencing Hormone Secretion," in *Behavioral Endocrinology*, ed. Jill B. Becker, S. Mark Breedlove and David Crews, Cambridge MA, MIT Press (1992): 401-422.
35. Paul C. Bernhardt, James M. Dabbs, Julie A. Fielden and Candice D. Lutter, "Testosterone Changes During Various Experiences of Winning and Losing Among Fans at Sporting Events," *Psychology and Behavior* 65 (1998): 59-62.
36. M.D. Matheson, and I.S. Bernstein, "Grooming, social bonding and agonistic aiding in rhesus monkeys," *Am. Jor. of Primatology* 51 (2000): 177- 186.
37. Randy Nelson, *An Introduction to Behavioral Endocrinology* 3rd ed, Sunderlan MA,Sinauer Associates 2005: 199f.
38. Dominic D.P. Johnson, Rose McDermott, Emily S. Barrett, Jonathan Cowden, Richard Wrangham, et al., "Over confidence in war games: Experimental evidence on expectations. aggression, gender and testosterone," *Proc. of the Royal Society B,* 272 (2006): 2513-2520.

39. F.E. Purifoy and L.H. Koopmans, "Androstenedone, testosterone and free testosterone concentration in women of various occupations," *Social Biology* 26 (1979):179-188.
40. J.M. Coates and J. Herbert, "Endogenous steroids and financial risk taking on a London trading floor," *Proc. Nat. Acad. Sciences USA* 105 (2008): 6167-6172.
41. Seymour Levine, "Cortisol Changes Following Repeated Experiences with Parachute Training," in *Psychology of Stress: A Study of Coping Men,* ed. Holger Ursin, Eivind and Seymour Levine, New York, Academic Press 1978: 51-56
42. Helen Bateup, Alan Booth, Elizabeth Shirtcliff and Douglas Granger, "Testosterone, cortisol and woman's competition," *Evolution and Human Behavior* 23 (2003): 181-192.
43. Stephen B. Manuch, Jay R. Kaplan and Franuis E. Latrich, "Brain Serotonin and Aggressive Disposition in Human and Nonhuman Primates," in *Biology of Aggression,* ed. Randy J. Nelson, Oxford U.K, Oxford University Press (2006): 65-113.
44. Ibid.
45. Joel Wallman, "Serotonin and Impulsive Aggression, Not So Fast," www.hfg.org/html.pages.wallman3.htm.
46. Biology of Violence, www.molbio.prinston.edu/courses/mb472/projects/0011/biology.htm
49. Randy J. Nelson, *An Introduction to Behvioral Endocrinology*, Sunderland MA, Sinauer Associates 2005: 515.
50. Neal Simon, Athena Cologer-Clifford, Shi-fang Lu, Suzanne E. McKenna and Shan Hu, "Testosterone and its Metabolites modulate 5HT1a and 5HT1b agonist effects in intermale aggression," *Neuroscience and Behavioral Reviews* 23 (1998): 325-336.
51. Larry J. Silver, "Neurobiology of Aggression and Violence," *Am. Jor Psychiatry* 165 (2008): 429-442.
52. Miczak and Fish 2006.

Chapter 9

1. William Shakespeare, *Henry V*, Prologue toAct II.
2. Global Trends 2030: Alternative Worlds, National Intelligence Council, Washington DC 2012: 49.
3. John Morgan Allman, *Evolving Brains,* New York, Scientific American Library 2000: 55.
4. David Dobbs, "Beautiful Brains," *National Geographic*, October 2011: 37-59.
5. Robert E. Ornstein and Richard F. Thompson, *The Amazing Brain*, Boston, Houghton Mifflin 1984: 24-25.
6. Richard M. Restak, *The Brain*, New York, Bantam Books 1984: 136.
7. Ibid: 5.
8. Ornstein and Thompson: 30-31.
9. Dobbs 2011.
10. Coalition to Stop the use of Child Soldiers, "Child Soldiers Global Report 2008," www.childsoldierglobalreport.org (accessed February 1, 2013).
11. UN High Commission on Refugees, "Child Soldiers Global Report 2001 Iran,html.unhc.org/refworld/country,CSCOAL, INF,498805f02d,0. html (accessed February 10, 2013)
12. Ibid.
13. Iran Military Forum, "Using children to run across a mine field to clear the way for ground troops," www.indopedia.org/Iran-Iraq.war.html (accessed February 11, 2013).
14. Photos courtesy of Gruppo Video Fotografico and USAid
15. World Armed Forces Forum, "95,000 child soldiers killed in thee Iran-Iraq war," www.network54.com/Forum/211833/thread/ 1348866642/ last 1348914965/95,000+Irnian+child+ soldiers+killed+Iran+Iraq+war (accessed February 11, 2013).
16. Helene Cooper, "In Homeland, Liberia Native Finds Resilience Amid Horror," *New York Times*, October 20, 2014.
17. Child Soldiers in the Civil War, hppt://civilwarsage.com/child-soldiers-in- the-civil-war/ (accessed August 26, 2014).
18. Cate Lineberry, "The Boys of War," *New York Times*, October 4, 2011.
19. Benedict Carey, "Shooting in the Dark," *New York Times*, February

11, 2013.

20. Craig A. Anderson, Akiko Shibuya, Nobuko Ihori, Edward L. Swing, Brad J. Bushman, et. l, "Violent Video Game Effects on Aggression, Empathy, and Prosocial Behavior in Eastern and Western Cultures: A Meta-Analytic Review," *Psychological Bulletin* 136 (2010): 155-173.
21. Craig A. Anderson, "An update of the effects of playing violent games," *Journal of Adolescence,* 27 (2004): 113-122.
22. Carey 2013.
23. Anderson 2004.
24. quoted in Carey 2013..
25. Michael R. Ward, "Video Games and Crime," *Contemporary Economic Policy*, 29 (April 2011): 648-61

Chapter 10

1. David Hume, *A Treatise on Human Nature* (1739), Oxford, UK, Clarendon Press 1896: 685.
2. Donald Rumsfeld, *Known and Unknown: A Memoir*, New York, Sentinel 2011: 711.
3. Joesph LeDoux, *Synaptic Self: How Our BNrains Become Who We Are,* New York, Vuking 2002, 10-44
4. Ibid:103
5. Maro de Curtis and Dennis Pare, "The hinal cortex: a wall of inhibition between the neocortex and the hippocampus," *Progress in Neurobiology* 74 (2004): 101-110.
6. Dominic D.P. Johnson and James H. Fowler, "The Evolution of Overconfidence," *Nature* 477 (2011): 317–320.
7. Edward O. Wilson, *On Human Nature*, Cambridge MA, Harvard Univ. Press 1978: 247.
8. Winston Churchill, *The World Crisis*, New York, Scribner's, 1931: 739, quoted in
9. Bernard Brode, *Strategy in the Missile Age*, Princeton, NJ, Princeton Univ. Press 1951: 60.
10. Bob Woodward, *Plan of Attack*, New York, Simon and Schuster 2004: 21.Bob Woodward, *State of Denial: Bush at War, Part III*, Simon and Schuster 2006: 77.

11. Rumsfeld 2006: 81
12. Woodward 2004: 26,
13. Condoleezza Rice, "Iraq: Goals, Objectives, Strategy," 22 October 2002, (This document is no longer available on the internet. I last accessed it on November 8, 2012)
14. Woodward 2004: 247–249.
15 Ibid.: 249-250.
16. Woodard 2004: 299-300.
17. C.J. Shivers, "Secret Causalities of Iraq's Abandoned Chemical Weapons," *New York Times*, October 14, 2014.
18. Collin Powell, UN Security Council Presentation, February 5, 2002,www. cnn/2003/US/02/05/sprj.irq.powell.transcript (accessed November 9, 2012).
19. Bob Woodward, *State of Denial*, New York, Simon and Schuster 2006: 218.
20. Woodward 2004: 313.
21. Ibid.: 313–315.
22. Fred Kaplan, *The Insurgents: David Petraeus and the Plot to Change America's Way of War,* New York, Simon and Schuster 2013: 74-75.
23. Ibid: 76.
24. Ibid: 58-59.
25. Ibid: 77.
26. Michael S. Schmidt, "In Blow to Government, Sadr Followers Call for New Elections," *New York Times*, December 26, 2011.
27. Tim Arango, "Summit Ends Boycott of Iraqi Parliament but Crisis Remains," *New York Times*, January 29, 2012.
28. Tim Arango, "Iraq's Political Crisis Eases as Ministers Return to Government," *New York Times,* February 7, 2012.
29. Michael S. Schmidt, "Rising Strife Threatens Iraqi Stability," *New York Times*, January 22, 2012.
30. Iraqi Body Count Project, www.iraqibodycount.org/database/ (accessed July 27, 2014).

Chapter 11

1. Jane Goodall, *The Chimpanzees of Gombe: Patterns of Behavior,* Cambridge, MA, Harvard Univ. Press 1986: 360.
2. Franz de Waal, *Peacemaking among Primates,* Cambridge, MA, Harvard Univ. Press 1989: 45-44.
3. James J. McKenna, "Biosocial Functions of Grooming Behavior among the Common Indian Langur Monkey (*Presbytis entellus*)," *Am. Jor. Physical Anthr- pology* 48 (1978): 503–510.
4. Franz de Waal, "The Integration of Bonding and Social Bonding in Primates," *Quarterly Rev. Biology* 61 (1986): 459–479.
5. Ibid.
6. Ibid.
7. This account of the Robbers Cave Experiment is drawn from Muzafer Sherif, O.J. Harvey, B. Jack White, William R. Hood, and Carolyn W. Sherif, *The Robbers Cave Experiment*, Middletown, CT, Wesleyan Univ. Press 1988.
8. Margaret Macmillan, *Paris 1919: Six Months that Changed the World*, New York, Random House 2002: 184, 192.
9 Frank P. Chambers, *This Age of Conflict,* New York, Harcourt, Brace and World 1962: 109-110.
10. William L: Schirer, *Diary: The Journal of a Foreign Correspondent 1934–1941,* New York, A.A. Knoff 1962: 82–90.
11. Chambers 1962: 109-110.
12. quoted in Schirer 1962: 88..
13. John McCormick, *The European Union*, Boulder, CO, Westview Press 1999: 35.
14. Ibid: 31.
15. Timothy Aston Ash, "The Crisis of Europe: How the Union Came Together and Why It's Falling Apart," *Foreign Affairs* 93 (September– October 2012): 2–15.
16. Ibid.
17. Pavel Molchanov, "Coalition of the Willing: Northern Ireland's Peace Breakthrough," *Foreign Policy Association* 2007.
18. John F. Burns, "New Violence in Belfast May Be about More Than a Flag," *The New York Times,* January 18, 2013.
19. Ibid.
20. Ibid.
21. After Richard Holbrooke, *To End a War*, New York, Random

House 1998: 23.

22. Patrice C. McMahon and Jon Western, "The Death of Dayton: How to Stop Bosnia from Falling Apart," *Foreign Affairs* 88 (2009): 60–83.
23. Ibid.
24. Dennis Ross, *Statecraft and How to Restore America's Standing in the World,* New York, Farrar, Straus & Giroux 2007: 232.
25. George J. Mitchell, *Making Peace*, New York, Alfred A. Knopf 1999: 47.
26. Ross 2007:2289
27. Ibid: 220.
28. Ibid: 220, 223.
29. Mitchell 1999: 9f.

Chapter 12

1. Ragendra K. Pachauri, Nobel Lecture, December 10, 2007, Oslo, Norway,hppt://nobleprize.org/nobel_prizes/peace/ laureatres/2007/ipcc-lecture_ em.html (accessed July 18, 2012).
2. China's Statements upon UNCLOS Ratification, June 7, 1996 and August 25, 2006.
3. John King Fairbanks, *The United States and China* (4th ed.), Cambridge, MA, Harvard University Press 1983: 158-161
4. Orville Schnell and John Delury, *Wealth an Power: China's Long March to the Twenty-First Century,* New York, Random House , 2013, Kindel Edition.
5. Ib Ohlsson, Reprinted by permission of *Foreign Affairs*, November/ December 2012), Copyright 2012 by the Council on Foreign Affairs, Inc.: 71
6. Steven Lee Myers and Jane Perlez, "No Movement On Key Disputes as Clinton Meets With Chinese Leaders," *New York Times*, September 6, 2012.
7. The Secretary of Defense, *Sustaining US Global Leadership: Priorities for 21st Century Defense*, Washington, US Department of Defense, January 2012: 2.
8. Keith Bradsher, "South China Sea Tensions a Backdrop to Kerry's China Visit," *New York Times,* February 14, 2014.

9 Helene Cooper and Jane Perlez, "U.S. Sway in Asia is Impearled as China Challenges Alliances," *New York Times*, May 30, 2014.

10. Ibid.

11. Kenneth Lieberthal and Wang Jisi, *Addressing U.S.-China Strategic Distrust*, Washington, DC, John L. Thornton Center at Brookings 2012.

12. Wang Jisi, "China's Search for a Grand Strategy: A Rising Great Power Finds Its Way," *Foreign Affairs* 90 (March/April 2011): 68-79.

13. Andrew J. Nathan. and Andrew Scobell, "How China Sees America," *Foreign Affairs* 91 (September/October 2012): 32-47.

14. Reuters, "China will beef up its military budget by 12.7 percent this year, the government said on Friday, a return to double digit increases that will stir regional unrest," *Reuters*, March 4, 2011.

15. Jisi 2011.

16. David M. Herzenhorn, "In Crimea, Russia Moved to Throw off he Cloak of Defeat," *New York Times*, March 24, 2014.

17. Mary Elise Sarotte, "A Broken Promise? What the West Really Told Moscow About NATO Expansion," *Foreign Affairs*, vol. 93, no. 5 (September/October 2014): 90-97.

18. Andrew Osborn, "Putin: Collapse of the Soviet Union was the Greatest Catastrophe of the Century," *The Independent*, April 26, 2005, www.independent.co.uk/world/news/Putin- Collapse- of-the-Soviet-Union-was-catastrophe-of-the-century-6147493.html ()accessed March 31, 2014).

19. David Brooks, "Putin Can't Stop," *New York Times*, March 3, 2015

20. Kuke N. Vargas, talkradionews.com

21. Steven Lee Myers and Graham Bowley, "Bush Wins NATO Backing on Shield, but not on Ukraine and Georgia," *New York Times*, April 3, 2008.

22. 2001 Ukrainian Census

23. Neil MacFarquhare, "Most Russian Forces Now Out of Ukraine, Kiev Says," *New York Times*, September 10, 2014.

24. North Atlantic Treaty Organization, "NATO leaders pledge support to Ukraine at Wales Summit," September 4, 2014, www.nato.int/cns/en/ natohg/news_112459.htm (accessed October 29, 2014).

26. Quran, 33: 36.

27. John L. Esposito, *Islam and Politics* (4th ed.) Syracuse Univ. Press 1984, p. 316.
28. Ibid: 116f.
29. David D. Kirkpatrick and Mary El Sheikh, "Egypt Military Enlists Religion to Quell Ranks," *New York Times*, August 23, 2013.
30. Tom Johnson, "Sharia and Militancy," Council on Foreign Relations, November 30, 2010, cfr.org/religion-and-politics/sharia-militancy/p- 19155 (accessed January 21, 2013).
31. Charles River Editors, *The Islamic State of Iraq and Syria: The History id the ISIS/ISIL*, San Bernardino CA, Charles River Editors 2014: 43-44.
32. Ibid.
33. Scott Shane and Ben Hubbard, "ISIS Displaying a Deft Command of Varied Media," *New York Times*, August 30, 2014.
34. ISIS Annual Reports Reveal a Metrics-Driven Military Command," Institute for the Study of War, May 22, 2014, www.undewrstandingwar.org/ISWBackgrounder_ISIS_Annual_Reports_0[1].pdf (accessed September 5, 2014).

Chapter 13

1. Winston Churchill, *My Early Life:A Roving Commission*, London, Butterworth 1930: 232
2. Abraham Lincoln, First Inaugural Address 1861

Appendix

1. Theodosius Dobzhansky, "Nothing Makes Sense in Biology Except in the Light of Evolution," *Am. Biology Teacher* 35(1973): 125–129.
2. Charles Darwin, *The Origin of Species by Means of Natural Selection*, `London, John Murray (1859). Facsimile ed., Harvard Univ. Press 1964.
3. Herbert Spencer, *Principles of Biology*. New York, D. Appl & Co 1864: 454-455.
4. Johann (Gregor) Mendel, "Versuche über Pflanzen–hybriden

(Experiments in Plant-hybridization)," *Verh. Natur. Verenis Brünn* (1865): 3–57. Translation reprinted in *Classic Papers in Genetics*, ed. James A. Peters, Englewood Cliffs 1959: 1–19.

5. National Genome Research Institute, National Institutes of Health, www.genome.gov/multimedia/illustrations/Fact Sheet_DNA.pdf. (Accessed October 1. 2014).
6. Sewall Wright, "Evolution in Mendelian Genetics," *Genetics* 16 (1931): 97–159.
7. Mutoo Kimura and Tomoko Ohta, "The Average Number of Generations until Fixation of a Mutant Gene," *Genetics* 61 (1969) 761–771.
8. Andraes Meyer-Lindenberg, Carolyn B. Mervis, and Karen Faith Berman, "Neural Mechanisms in Williams Syndrome: A unique Window to Genetic Influences on Cognition and Behavior," *Nature Reviews Neuroscience* 7 (2006): 380–393.
9. Lucy R. Osborne, Martin Li, Barbra Pober, David Chitayat, Joannn Boduirtha et al., "A 10 Million-Base Pair Inversion Polymorphism in Families with Williams-Beuren Syndrome," *Nature Genetics* 28 (2001): 321–325.
10. Ursula Bellugi, Loz Lichtenberger, Debra Mills, Albert Galaburda, 0and Julie. Korenberg, "Bridging Cognition, the Brain and Molecular Genetics from Williams Wyndrome," *Trends in Neuroscience* 22 (1999): 97–107.
11 Liangbiao Chen, Arthus L. DeVries, and Chi-Hing C. Cheng, "Convergent Evolution of Glycoproteins in Artic Notothenioid and Artic Cod," *Proc. Nat. Academy of Science USA* 94 (1997): 3817–3822.
12. Niles Eldredge and Stephen Jay Gould, "Punctuated Equilibria: An Alternative to Phyletic Gradualism," in *Models in Paleobiology*, ed. T.J.M. Schopf, San Francisco, Freeman, Cooper and Co. 1972: 82–115.
13. Mary-Claire King, and Allan Wilson, "Evolution at Two Levels in Humans and Chimpanzees," *Science* 188 (1975): 107–16.

Persons Index

Subject Index

CPSIA information can be obtained
at www.ICGtesting.com
Printed in the USA
FFOW02n1907020216
21052FF

9 781457 535482